# Prosthetic Biomechanics in Engineering

T0094166

# Prosthetic Biomechanics in Engineering

Edited by
N A Abu Osman

## CRC Press
Taylor & Francis Group
Boca Raton  London  New York

CRC Press is an imprint of the
Taylor & Francis Group, an **informa** business

First edition published 2022
by CRC Press
6000 Broken Sound Parkway NW, Suite 300, Boca Raton, FL 33487-2742

and by CRC Press
2 Park Square, Milton Park, Abingdon, Oxon, OX14 4RN

Library of Congress Cataloging-in-Publication Data
Names: Abu Osman, N. A. (Noor Azuan), editor.
Title: Prosthetic biomechanics in engineering / edited by N.A. Abu Osman.
Description: Boca Raton : CRC Press, 2022. | Includes bibliographical
references and index. | Summary: "Prosthetic Biomechanics in Engineering
is about the recent advances in prosthetic engineering research. The
scope of the book is focused on the design, development and evaluation
of a prosthetic systems that are being used in biomechanical
applications"-- Provided by publisher.
Identifiers: LCCN 2021034730 (print) | LCCN 2021034731 (ebook) | ISBN
9781032052410 (hardback) | ISBN 9781032052458 (paperback) | ISBN
9781003196730 (ebook)
Subjects: MESH: Prosthesis Design--methods | Biomechanical Phenomena |
Biocompatible Materials | Bioengineering--methods
Classification: LCC R857.M3 (print) | LCC R857.M3 (ebook) | NLM QT 37 |
DDC 610.28--dc23
LC record available at https://lccn.loc.gov/2021034730
LC ebook record available at https://lccn.loc.gov/2021034731

ISBN: 978-1-032-05241-0 (hbk)
ISBN: 978-1-032-05245-8 (pbk)
ISBN: 978-1-003-19673-0 (ebk)

DOI: 10.1201/9781003196730

Typeset in Times
by SPi Technologies India Pvt Ltd (Straive)

# Contents

# Notes on the Editor

Dr Noor Azuan Abu Osman graduated from University of Bradford, UK with a BEng Hons in mechanical engineering, followed by MSc and PhD in bioengineering from University of Strathclyde, UK. He is a practicing engineer and Professor of biomechanics with the Department of Biomedical Engineering, Faculty of Engineering, University of Malaya, Malaysia. His research interests include measurements of human movement, prosthetics design, the development of instrumentation for forces and joint motion, and the design of prosthetics, orthotics and orthopaedics. Prior to joining University of Malaya in 1995, he worked as a Mechanical and Electrical Engineer and is actively involved in many consultancy projects, especially in the field of biomechanics and biomechanical engineering.

# 1 Fibre Bragg Grating (FBG) Sensor for Socket Pressure Measurement

## N A Abu Osman
University of Malaya, Kuala Lumpur, Malaysia

## E A Al-Fakih
Imam Abdulrahman Bin Faisal University, Dammam, Saudi Arabia

## F R Mahamd Adikan
University of Malaya, Kuala Lumpur, Malaysia

## CONTENTS

DOI: 10.1201/9781003196730-1

## 1.1   INTRODUCTION

### 1.1.1   BACKGROUND

In Malaysia, a series of National Health and Morbidity Surveys (NHMS) reported that the prevalence of diabetes mellitus (PVD) among Malaysian adults was 6.3% in 1986, 8.3% in 1996, and 11.6% in 2006 (Letchuman et al., 2010). In the latest NHMS report issued in 2015 by the Ministry of Health Malaysia, the prevalence had shockingly increased to 17.5% (NHMS, 2015).

### 1.1.2   BELOW-KNEE PROSTHESIS

Below-knee amputees usually use a prosthesis as a rehabilitation device to restore their lost limb for daily activities (Al-Fakih et al., 2016a). A prosthesis includes few crucial parts, such as a foot, pylon, socket and ankle. The socket comes up with the coupling between the remaining parts of the prosthetic tool and residual limb. The design and fitting of socket are the most difficult procedures because of the complexity and distinctness of each amputee's residual limb (Laing et al., 2011). This kind of transfer can be achieved with a high mastery in the interface pressure issue within prosthetic sockets because it has a major effect on socket comfort and quality fit (Laing et al., 2011).

### 1.1.3   INTERFACE PRESSURE DISTRIBUTION WITHIN BELOW-KNEE SOCKETS

The stereotypical types of sockets for below-knee amputees are the total surface bearing (TSB) and patellar tendon bearing (PTB) (Laing et al., 2011). The PTB socket idea issue loads over pressure-tolerant areas of the residual limb such as the patellar tendon (PT), medial flare of tibia, anterior muscular compartment and popliteal area, while pressure is proved on intolerant areas, such as the anterior distal tibia, fibula head and anterior tibia crest (Baars and Geertzen, 2005; Safari and Meier, 2015). The sockets of TSB, residual limb soft tissues are exposed to bearable higher pressure, and the bony areas are stabilized in the residual limb (Laing et al., 2011); thus, huge loads can escape from skin damage occurs when the uses of silicone liners (Eshraghi et al., 2013).

### 1.1.4 PROBLEM STATEMENT

PTB sockets show an indentation at the PT region, which bears the highest pressure from the total body weight (Abu Osman et al., 2010a). TSB socket is also character-ized by its ability was covered the loads all over the residual limb evenly. Almost all below-knee amputees still complain that their prosthesis causes various complica-tions because of the misdistribution of interface pressure and shear stresses within the prosthetic sockets (Dou et al., 2006; Reiber, 1994; Sengeh and Herr, 2013).

Therefore, many researchers have developed various interface stress transducers in the past 50 years to quantify these interface stresses in recognize areas with higher stresses that may produces skin breakdown, stress covered area in compared in vari-ous socket designs, and evaluate suspension systems and interface cushioning mate-rials (Ali et al., 2012; Polliack et al., 2000). The uses in transducers calculated as helped prosthetists or researchers appreciated the mechanical interactions occurring between the socket and residual limb.

The most common measurement techniques are strain gauge (SG)-based trans-ducers, in-socket sensor mats (e.g., F-socket transducers), and finite element analysis (FEA) methods (Abu Osman et al., 2010b; Dumbleton et al., 2009; Moo et al., 2009). The bulky size of SG-based sensors and their data cables increases weight in the prosthesis, thereby distorting the stress measurement (Zhang et al., 1998).

## 1.2 THE EVOLUTION OF BELOW-KNEE SOCKET DESIGNS

The artificial joint-corset below-knee prosthesis which practices the thigh corset, the uses or not a waist belt, as a suspension technique had been used numerous years before Radcliffe proposed the patellar tendon bearing (PTB) sockets in the 1950s (Radcliffe, 1962; Radcliffe and Foort, 1961; Wu et al., 2003). The PTB socket is stereotypically create in laminated woven materials together with acrylic resins or of moulded thermoplastic sheets (Ng et al., 2002). The structured socket delivers cov-ered of the patella tendon (the distal third of the patella) and extends the medial and lateral aspects of the socket higher up to the level of adductor tubercle of the femur in order that it shares body weight bearing and secured stability in knee.

The PTB architecture issues loads over pressure-tolerant areas of the residual limb, such as the patellar tendon (PT), the anterior muscular compartment, the medial flare of the tibia, and the popliteal area, while relieving pressure on intolerant areas, such as the fibula head, anterior tibia crest, and anterior distal tibia (Baars and Geertzen, 2005; Safari and Meier, 2015). This theory had proven satisfactory out-comes for up to 90% of amputees (Galdik, 1955) and still one of the most used socket types (Friel, 2005). PTB socket fabrication is a time-consuming and labour-intensive process that requires a professional prosthetist (Abu Osman et al., 2010a).

The absolute adhesion of the silicone liner to the residual limb, preservation of the residual limb, better aesthetic appearance, and increased function make TSB sockets superior suspension (Baars and Geertzen, 2005). Amputees have stated a preference with prosthetic sockets incorporating silicone liners as suspension systems (Eshraghi et al., 2012). Recently, new suspension systems (air pneumatic (Pirouzi et al., 2014) and magnetic systems (Eshraghi et al., 2012)) have been proposed, and it shown to be able to transcend the disadvantages of current schemes in terms of volume fluctuations over time, donning and doffing, and the discomfort caused by the milking phenomena.

## 1.3  FIBRE BRAGG GRATING SENSORS' CAPABILITY TO MEASURE SOCKET INTERFACE PRESSURES

### 1.3.1  PTB SOCKET FABRICATION

The major flows in manufacturing PTB sockets are the following: first, wrapping plaster of Paris (POP) around the soft tissue of the residual limb creates a negative mould of the residual limb. The POP is left for a while to allow the residual limb to take on its proper shape. Second, the positive mould of the residual limb is produced based on the shape of the negative cast. Third, to achieve the desired shape, the positive mould is adjusted or rectified. Fourth, the socket liner, which is made of polyethylene, is placed together. After that, the product is laminated. Socket finishing is the final move (Abu Osman et al., 2010a).

### 1.3.2  FBG FABRICATION TECHNIQUE

A constant FBG was constructed using the phase mask practice. The phase mask is a flat slab of silica glass that is transparent to UV light. The optical fibre is almost in contact with the corrugations of the phase mask, and UV light incident in the usual direction passes through the mask and diffracts in the 0, +1, and 1 diffracted order by those corrugations. For this analysis, a 10mm long FBG with a Bragg peak wavelength of around 1,550.952nm was developed.

### 1.3.3  SENSOR DESIGN AND FABRICATION

The application required advanced protection. Epoxy used to protect the FBG fibre (Table 1.1). After that, the epoxy pad was sandwiched between two sheets of silicone polymeric materials, which served as the pressure sensor (Figure 1.1).

**TABLE 1.1**
**Typical Properties of NOA61**

| Solids | 100% |
|---|---|
| Viscosity at Room Temperature | 300 cps |
| Refractive Index of Cured Polymer | 1.56 |
| Elongation at Failure | 38% |
| Modulus of Elasticity (psi) | 150,000 |
| Tensile Strength (psi) | 3,000 |
| Hardness – Shore D | 85 |

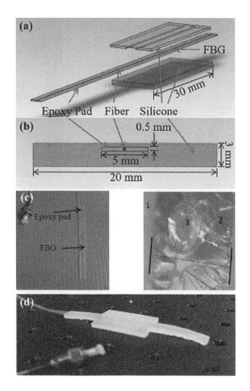

**FIGURE 1.1** Assembly of the pressure instrument. The epoxy pad (the strip-like pad indicated in (a)) was positioned between the upper (1mm thick) and the lower (2mm) layers of silicone materials to form the pressure sensor; (b) The sensor cross-sectional area is 20 × 3mm² and the dimensions of the whole sensor were 30 × 20 × 3mm³, and the sensitive surface area was 20 × 30mm²; (c) Images of the epoxy pad and its microscopic cross-sectional area; (d) The image of the whole pressure sensor.

### 1.3.4 SENSOR ADJUSTMENT

The sensor adjusted by utilizing an Instron Microtester 5848 stress machine where mostly pressure concentrated to imitate this behaviour by utilizing a ball bearing for the adjustment to utilize similar concentrated compressive packs onto the sensor (Figure 1.2).

### 1.3.5 EXPERIMENTS

At the PT bar area, the sensor assembly was inserted into the below-knee socket (Figure 1.3). The PT bar was aligned using Abu Osman et al. (2010a) to ensure that the sensor installation was in the correct location. For the amputee's comfort and to reduce the risk of damage caused by the residual limb pistoning inside the socket, the sensor was positioned flush with the inner socket wall.

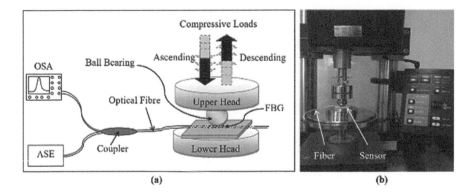

(a)                                              (b)

**FIGURE 1.2**  Representation of the sensor adjustment. (a) Simple chart diagram proving the calibration experimental system; (b) Image depicting the system.

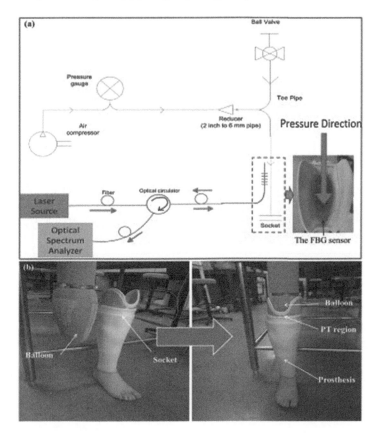

**FIGURE 1.3**  Trial system of the sensor functionality trial. (a) A diagram of the *in situ* test, showing the sensor while connected to OSA and subjected to pressure loads. The inset image shows the sensor/socket integration keeping the sensor sheet flush with the socket liner and the red-coloured thick arrow illustrates the vector of pressure application in perpendicular to the FBG sensor; (b) Images depicting the setup and the heavy-duty balloon used for this study.

---

**TABLE 1.2**
**The Details of FBG Sensor**

| Specifications | Values |
| --- | --- |
| Dynamic Range (N) | 30 |
| Maximum WL Shift (pm) | 4,000 |
| Sensitivity (pm/N) | 127 |
| Expected error width (nm) | 0.34 |
| Hysteresis (FSO) | 9% |
| Maximum FBG Disp. (mm) | 1.39 |
| Maximum Power Drop (dB) | −28.9 |
| Alignment of FBG | Properly aligned |

---

### 1.3.6 Results and Discussions

The sensor goes through seven trials. The average values were used to interpret the sensor behaviour, which indicated that the FBG sensor was sensitive to small changes in the applied forces. Table 1.2 provide details of the manufacturing the FBG sensor.

The sensor revealed a linear reaction which agrees with the optical and mechanical characteristics of FBGs (Luyckx et al., 2010). The line crossing with the y-axis at 0.0244 (nm) is nearly zero when compared with the sensor FSO, which is about 3.8nm (~0.0065 of FSO). The sensor revealed a very tolerable FSO hysteresis error of roughly 0.09.

While the FBG was embedded, the curing phase of epoxy and silicone-polymers was controlled. The host materials develop curing, causing in a minor negative shift in the FBG reflected wavelength (Figure 1.4(a) and (b)).

The FBG sensor is a versatile pressure sensor that can be used to measure pressure at any irregularly formed interface, such as the residual limb-socket interface. Involving patients as research subjects necessitates careful consideration of temperature fluctuations, which necessitates the isolation of the sensor.

## 1.4 DEVELOPMENT AND VALIDATION OF FIBRE BRAGG GRATING SENSING PAD FOR INTERFACE PRESSURE MEASUREMENTS WITHIN PROSTHETIC SOCKETS

### 1.4.1 Sensing Pad Fabrication and Experimental Setup

Two aluminium plates were used to make two custom moulds that allowed for the development of sensing pads of various thicknesses and FBG fibre embedding depths (Figure 1.5). To hold the FBG fibres in place during fabrication, they were put in their intended positions and fixed firmly at the mould edges in specially engraved grooves. The silicone polymeric materials used in fabrication were made by combining a copolymer and its catalyst in a specific ratio that could be modified depending on the

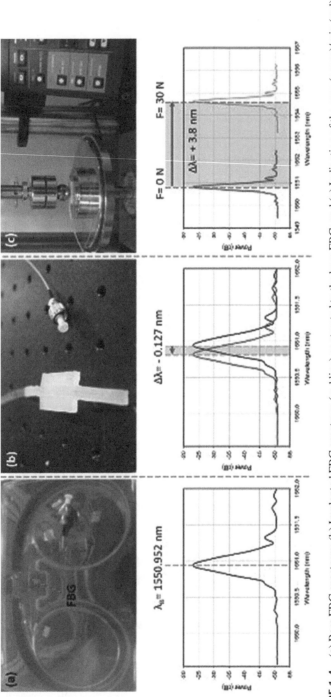

**FIGURE 1.4** (a) Bare FBG spectrum; (b) Implanted FBG spectrum (red line) contrasted to the bare FBG; and (c) Indication of the max (green)/min (red) shifts caused by the max/min applied forces during calibration.

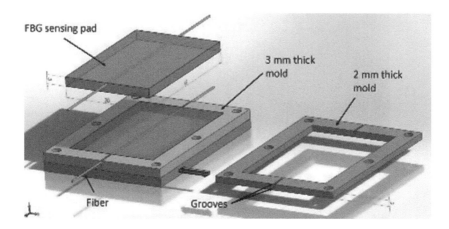

**FIGURE 1.5** Mould design and sensing pad fabrication. Two moulds with thicknesses of 2mm and 3mm are custom made.

material type (10:1 by weight for the PDMS). The mixture was then poured into the mould and allowed to cure for 24 hours.

The first sensing pad's top layer and the neutral layer sensing pad's other layer (Figure 1.6(a)). The top position arrangement was set upside down to perform as if the FBG was located at the bottom layer of the sensing pad (Figure 1.6(a, right)). The three FBG arrangement was loaded under three various measurement settings: bare

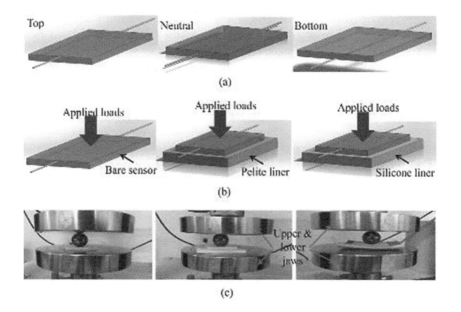

**FIGURE 1.6** (a) Three different FBG arrangements (top, neutral and bottom); (b) FBG conditions: no liner, attached to the pelite and silicone liners (left to right); (c) Actual images of the three different conditions.

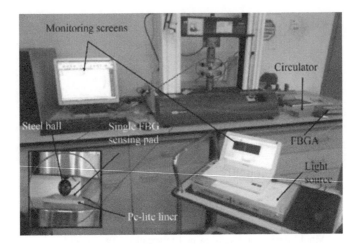

**FIGURE 1.7**    The trials for the mechanical testing of the FBGs different arrangements.

sensing pad, attached to the pelite, and to silicone liners (Figure 1.6(b)). This allowed us to compare the sensitivity of all three settings for each measurement condition separately, allowing us to determine which settings provides the most sensitive pad for each measurement condition. Figure 1.6(c) reveals actual images of the "top position" sensing pads under the three various measurement settings.

Every sensing pad's FBG fibre pigtail was connected to a broadband light source, and the reflected Bragg wavelength was measured with the Fibre Bragg Grating Analyzer. A steel ball was attached to the machine's upper jaw to add loads resembling those found at the residual limb-socket interface of below-knee sockets in real life. (Figure 1.7).

The efficiency comparison was limited to sensing pads with thicknesses of 2mm and 3mm. The trials were continued by repeated to evaluate the sensitivity of the two 2mm and 3mm thick sensing pads under the various environments: attached to the pelite liner, silicone liners and no prosthetic liner.

One of the key fabrication parameters influencing FBG sensitivity and durability was also thought to be the form of sensing pad materials. The first FBG was inserted in the PDMS (harder), and the second in the silicone rubber (softer). The fabrication procedures for silicone rubber sensing pads were similar to those for PDMS, with the exception of a 20:1 weight ratio between the liquid base and curing agent.

Following that, four new expandable sensing pads with at least two sensing sites were created to cover up the distal and proximal subareas of each socket wall (lateral, medial, anterior and posterior) and were characterized with a sensitivity of 60 pm/kPa (Figure 1.8).

The amputee's residual limb was then fitted with sensing pads, and he stepped on a treadmill. A LabVIEW software was used to only pick up the coordinates of the reflected spectra's peak wavelengths at 50 times/s for each FBG. Figure 1.9 depicts the sensing pads, F-socket mats, and the subject while walking on the treadmill.

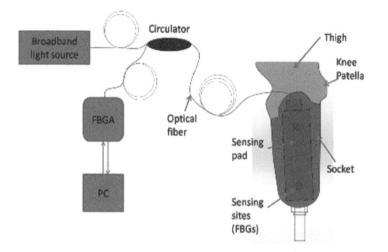

FIGURE 1.8   Typical outline of the trial for pressure measurement in eight subareas of the residual limb.

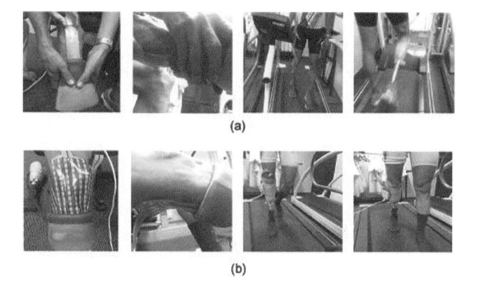

FIGURE 1.9   The fitting of the pressure sensors in the socket and the subject while walking on the treadmill. (a) and (b) demonstrate the arrangement of the FBG sensing pads and the F-socket transducers.

### 1.4.2  FBG's Embedding Depth

Figure 1.10 and Figure 1.11 compare the sensitivity of FBGs embedded in the top, neutral, and bottom layers of 3mm thick PDMS sensing pads under three different environments: no liner, fastened to the pelite liner, and fastened to the silicone liner. All of the FBG sensors demonstrated good linearity and measurement precision, which is consistent with their optical properties.

### 1.4.3  Type/Hardness of Sensing Pad Materials

The silicone rubber substance "Shore A" with a lower hardness of 13 was used. Figure 1.12 depicts a comparison of the effects of FBG embedment in PDMS (harder) and silicone rubber (softer) host materials on sensing pad output under three environments: bare sensing pad, fastened to the pelite liners or to silicone liners. When attached to both types of socket liners, the harder host material (PDMS) produces less hysteresis error than the softer silicone rubber (Figure 1.13).

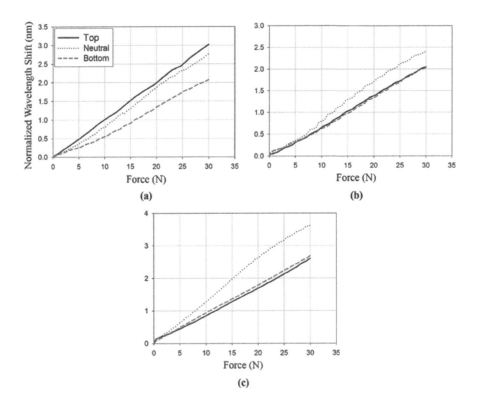

**FIGURE 1.10**  Sensitivity evaluation of three different FBG arrangements loaded under three different environments. (a) Bare pad (no prosthetic liner); (b) by the pelite liner; (c) by the silicone liner.

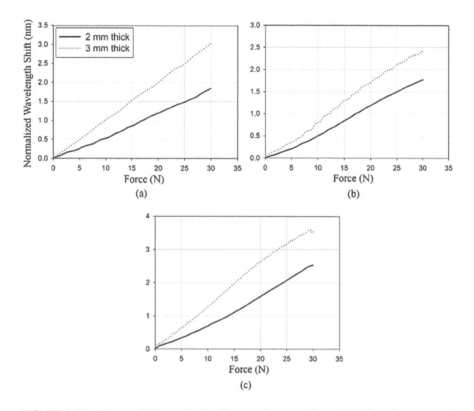

**FIGURE 1.11** The sensitivity evaluation between 2mm and 3mm PDMS sensing pads with the FBGs implanted at the neutral layer. (a) Bare sensing pad; (b) fastened by pelite liner; (c) fastened by silicone liners.

When subjected to cyclic loads, both the PDMS and silicone rubber host materials demonstrated excellent consistency (Figure 1.14). The hysteresis error was greater in the softer (silicone rubber) host material than in the harder host materials.

### 1.4.4 INTERFACE PRESSURE MAPPING AND SENSOR VALIDATION

For testing inside the prosthesis, a 3mm expandable sensing pad with an array of two FBGs embedded in the neutral layer of the PDMS was used. Figure 1.15 depicts pressure data collected during amputee gait at the PT area.

For contrast, pressure measurements were taken at eight subregions of the amputees' residual limbs using the FBG sensors and F-socket mats. Table 1.3 shows the mean peak pressure values produced by these two techniques. Figure 1.16 depicts the peak pressure trends at all subregions. The maximum pressure (152kPa) was measured at the residual limb's anterior proximal surface. This corresponded to the pressure values measured during normal walking in our previous research. According to another report, the average peak pressure was less than 200kPa. When compared to the F-socket mats, the FBG sensors reported higher pressure values. This disparity

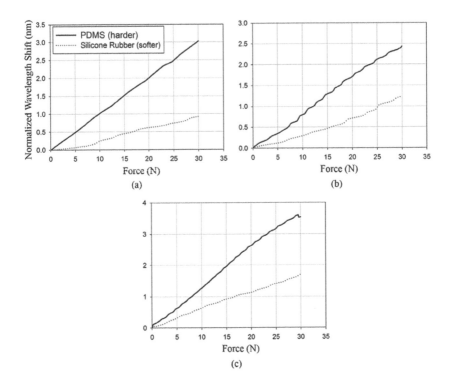

**FIGURE 1.12** FBGs response when implanted in the neutral layer of PDMS (harder) and silicone rubber (softer) sensing pads. (a) Bare sensing pad; (b) fastened to the pelite liner; and (c) fastened to the silicone liner.

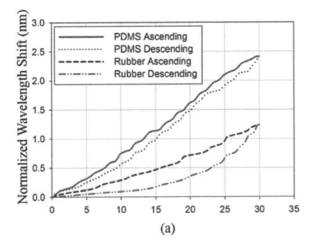

**FIGURE 1.13** The hysteresis error of the FBGs implanted in the neutral layer of harder (PDMS) and softer (silicone rubber) host materials. The error when fastened to (a) Pelite liner.

*(Continued)*

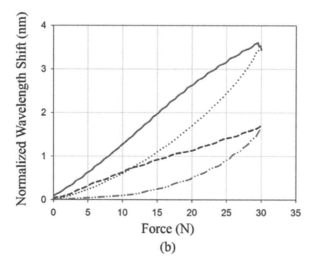

**FIGURE 1.13 (Continued)**    (b) silicone liners.

may be due to the thickness of the FBG sensors (3mm) versus the F-socket sensing mats (0.2mm). The thicker sensor may have raised the overall pressure at the sensing sites because the socket size was the same. Nonetheless, the pressure shift pattern has been consistent for both sensors.

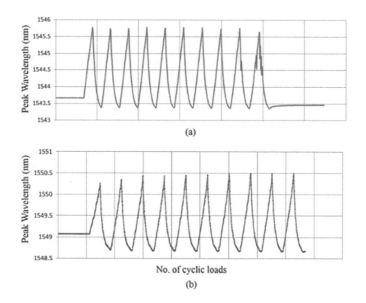

**FIGURE 1.14**    The repeatability trial. The repeatability of FBG sensors when implanted in (a) PDMS, and (b) silicone rubber host materials.

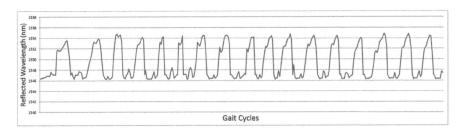

**FIGURE 1.15** The pressure verified at the patellar tendon during repetitive gait cycles.

**TABLE 1.3**

**The Average Peak Pressures Verified at 8 Detection Points Using FBG Sensors and F-Socket Transducers**

| Residual Limb Area | FBG Sensors | F-Socket Transducers |
|---|---|---|
| Anterior proximal | 118.44 | 109.74 |
| Anterior distal | 48.63 | 41.82 |
| Posterior proximal | 48.37 | 41.01 |
| Posterior distal | 51.42 | 46.17 |
| Lateral proximal | 56.25 | 43.04 |
| Lateral distal | 55.19 | 39.28 |
| Medial proximal | 63.97 | 59.37 |
| Medial distal | 38.14 | 35.86 |

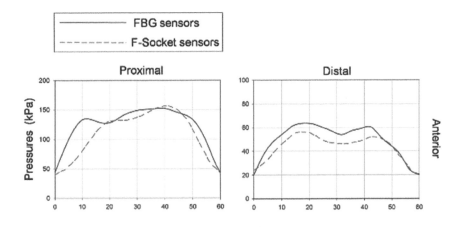

**FIGURE 1.16** The peak pressures at eight nominated subareas of the amputee socket using FBG sensing pads and F-socket transducers.

*(Continued)*

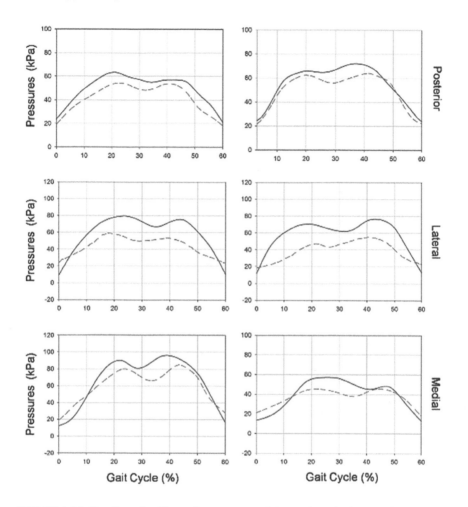

**FIGURE 1.16 (Continued)**    The peak pressures at eight nominated subareas of the amputee socket using FBG sensing pads and F-socket transducers.

## 1.5  SILICONE LINER WITH FIBER BRAGG GRATING FOR PRESSURE MEASUREMENT IN BELOW-KNEE AMPUTEES' PROSTHETIC SOCKETS

### 1.5.1  Silicone Sensing Liner Fabrication, Mould Fabrication

The inner solid mould was created by a gypsum cone-shaped positive cast (Figure 1.17(a)). To keep the requisite space between the inner mould and the outer shell, a commercially available silicone liner was rolled over the inner mould, followed by an additional layer of stockinet (Figure 1.17 (a, right)). These shells were joined together as the assembly progressed to form a conical-shaped outer shell reveals in Figure 1.17 (b, left). The additional plastic was trimmed, and the vacuum was kept on during the process. The two shell halves were left in place until fully cold. Before disassembling

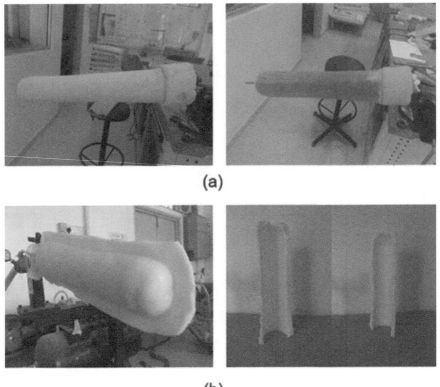

**FIGURE 1.17** Fabrication of mould. (a) Interior mould production and revision, and (b) exterior shell manufactured of two halves.

the shells, a few holes were drilled along the sides to allow the two halves to be reassembled while fabricating the FBG-instrumented silicone liner (Figure 1.17 (b, right)).

## 1.5.2 CUSTOM FABRICATION OF FBG-INSTRUMENTED SENSING LINER

First, a 1mm thick silicone layer was formed by uniformly brushing silicone with shore hardness A 20 directly over the surface of the inner gypsum mould (Figure 1.18 (a)). In locations of clinical interest, such as the proximal, central, and distal regions of all four aspects (anterior, posterior, medial and lateral) of the limb mould, a total of 12 FBG elements were positioned in a special arrangement over the inner silicone layer (Figure 1.18(b)). Each of these FBGs was previously crammed between two thin layers of stockinet for highest safety, as shown in the inset of Figure 1.18(b). This was accomplished by assembling the outer shell's two halves over the first inner layer of silicone and FBG sensing pads (Figure 1.18(c)). Guidance grooves were cut out to secure the outer shell and retain the vital space between the inner silicone layer and shell. After disassembling the outer casing, the FBG-instrumented liner was ready for use (Figure 1.18(d and e)).

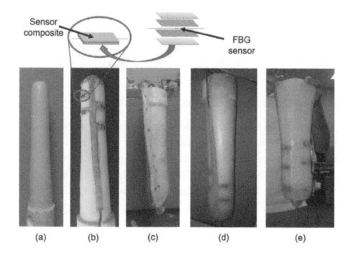

**FIGURE 1.18** Sensing silicone liner fabrication.

### 1.5.3 SENSOR LINER CALIBRATION

For the sensing liner calibration, a unique experimental setup was developed (Figure 1.19). The sensing liner was calibrated to imitate the actual socket interface condition. A heavy-duty balloon with a 1mm thick wall was manufactured and then inserted into the sensing liner, which was then mounted inside a conical-shaped hollow mould made specifically for this setup, as shown in Figure 1.19. A PVC pipe was

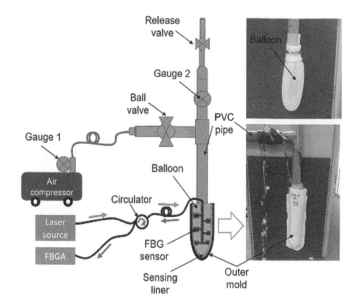

**FIGURE 1.19** Setting stage for the FBG-instrumented liner. The two insets reveal the fabricated balloon and the exterior mould used for the tuning process.

then used to link the balloon to an air compressor. A pressure gauge was installed in the pipe to show the amount of pressure supplied to the balloon. The fibre pigtail of each of the 12 FBG sensing components was coupled into a broadband light source, and the FBGA was used to calculate the change in the reflected Bragg wavelengths.

### 1.5.4   BELOW-KNEE PROSTHETIC SOCKET FABRICATION

For this experiment, a TSB socket was developed. The following are the key steps in the fabrication of TSB sockets. First, the POP rolls were used to create a negative image of an amputee's residual limb. Second, the negative cast was used to build the positive mould of the residual limb. Third, the positive mould was altered to obtain the final form of the socket. Finally, a 12mm thick Northplex plastic sheet was heated in the oven for around 20 minutes at 150°C. It was then draped over the positive mould and allowed to cool for a few minutes. The extra plastic was trimmed so that the socket matched the outline of the planned residual limb.

### 1.5.5   ARTIFICIAL RESIDUAL LIMB FABRICATION

Since no subjects took part in this research, attempts were made to simulate a real-life scenario. A replica of the residual limb was made. The real residual limb is made up of bones (upper tibia and fibula), muscles and skin. Plastic polyester resin was used to create a copy of the upper parts of the tibia and fibula bones (Figure 1.20(a)). The skins and muscles were made of silicone with shore A hardness of 20 and 10, respectively. The final limb model (Figure 1.20(b)).

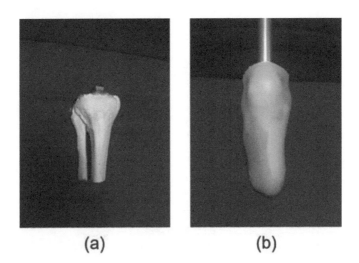

**(a)**                                    **(b)**

**FIGURE 1.20**   Fabrication of artificial residual limb. (a) A model of the upper portions of the bones; tibia and fibula; (b) Final synthetic residual limb. The brown silicone is the muscle tissues while the white soft silicone is the skins.

### 1.5.6 EXPERIMENTAL SETUP AND PROCEDURES

Figure 1.21(a) shows the FBG-instrumented liner being rolled over the artificial residual limb. Commercially available F-socket mats were also used for pressure measurements at the same sensing areas in order to evaluate the results obtained using the two sensing methods at the same time (Figure 1.21(b)). The sensor mats were cut to suit the outlines of the residual limb and placed on the anterior, posterior, medial and lateral sides. After that, the prosthetic was securely connected to the gait simulating system. Figure 1.21(c) depicts the FBG-instrumented liner and F-socket mats being installed in the prosthesis.

The pressure profile was calculated as a function of the wavelength shifts measured. For each FBG, a LabVIEW software capable of logging only the coordinates of the peak wavelengths of the reflected spectra at a sampling frequency of 50 Hz was

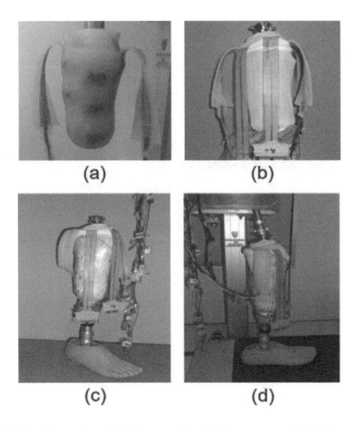

(a)

(b)

(c)

(d)

**FIGURE 1.21** The setting trials for measuring device pressure with FBG sensors and F-socket mats. (a) The liner was rolled over the artificial residual limb with the FBG-instrumented liner; (b) The liner was then coated with a thin stockinet and four F-socket mats were adhered to it; (c) and (d) demonstrate the incorporation of both sensing techniques into the prosthesis assembly.

*(Continued)*

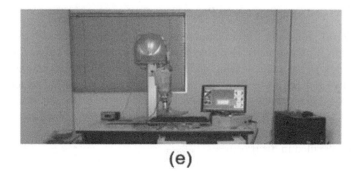

(e)

**FIGURE 1.21 (Continued)**   (e) The prosthesis assembly is attached to the gait simulating system, which is operational.

developed. Likewise, four F-socket mats covering all socket aspects were mounted over the FBG-instrumented liner, and their ends were attached to the Tekscan system. (Figure 1.21(d)).

### 1.5.7   LINER CHARACTERIZATION

Figure 1.22 depicts, with regression lines, the calibration curves and standard deviations of all 12 FBG sensing components. Table 1.4 summarized the sensitivities, calibration equations and calibration dynamic ranges of the sensors. All of the FBG sensors have clearly shown a linear relationship between the changes in the back-reflected wavelengths and the applied pressures, which is consistent with the mechanical and optical properties of FBGs (Al-Fakih et al., 2016b).

### 1.5.8   INTERFACE PRESSURE PROFILE USING THE TWO MEASUREMENT TECHNIQUES

The interface peak pressures measured with the newly developed FBG sensing liner were compared to the pressure profiles measured by the widely used F-socket sensing mats when walking. The data obtained from the simulating machine's 10 cycles of gait simulation were statistically analyzed using the t-test. The mean peak pressures for different areas of the residual limb were contrasted, and p-values were less than 0.05 (Table 1.5), indicating that both mapping methods computed pressure in a consistent and non-significant manner.

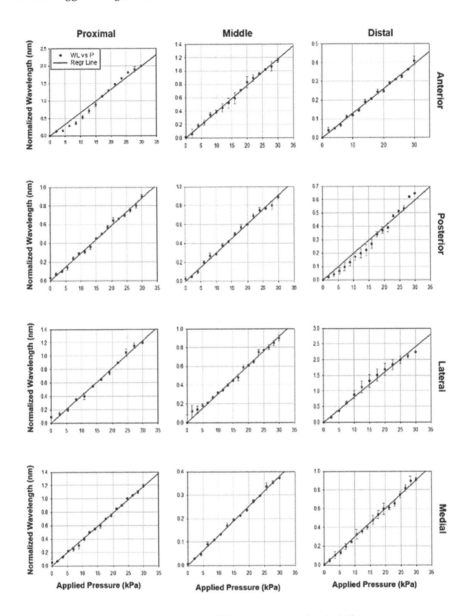

**FIGURE 1.22** The tuning curves of 12 FBG sensors screening their linear response.

**TABLE 1.4**

**Sensitivities, Dynamic Ranges, and Calibration Equations for the Twelve FBG Sensors**

| | | | Calibration Dynamic Range | | | |
|---|---|---|---|---|---|---|
| Region | Sensor site | Sensitivity (kPa/nm) | Min. WL (nm) | Max. WL (nm) | Max. WL shift (nm) | Equations |
| Anterior | Proximal | 15.12 | 1,550.23 | 1,552.22 | 1.99 | P = 15.12 WL |
| | Middle | 25.51 | 1,550.19 | 1,551.35 | 1.16 | P = 25.51 WL |
| | Distal | 76.69 | 1,539.95 | 1,540.36 | 0.41 | P = 76.69 WL |
| Posterior | Proxima | 34.12 | 1,550.31 | 1,551.21 | 0.90 | P = 34.12 WL |
| | Middle | 30.11 | 1,554.92 | 1,555.81 | 0.89 | P = 30.11 WL |
| | Distal | 50.53 | 1,556.12 | 1,556.77 | 0.65 | P = 50.53 WL |
| Lateral | Proximal | 24.45 | 1,549.72 | 1,550.92 | 1.20 | P = 24.45 WL |
| | Middle | 32.76 | 1,547.23 | 1,548.13 | 0.90 | P = 32.76 WL |
| | Distal | 12.46 | 1,549.55 | 1,551.79 | 2.24 | P = 12.46 WL |
| Medial | Proximal | 25.41 | 1,555.79 | 1,556.98 | 1.19 | P = 25.41 WL |
| | Middle | 79.10 | 1,555.11 | 1,555.48 | 0.37 | P = 79.10 WL |
| | Distal | 33.18 | 1,550.09 | 1,551.00 | 0.91 | P = 33.18 WL |

**TABLE 1.5**

**During the Stance Process, the Mean Peak Pressure Values (kPa) at the Residual Limbs Regions**

| | | Mean Peak Pressure (SD) | | |
|---|---|---|---|---|
| Region | Subregion | FBG Sensors | F-Socket Sensor | P-Value |
| Anterior | Proximal | 17.23 (1.23) | 16.52 (0.99) | 0.05 |
| | Middle | 34.18 (4.80) | 35.56 (4.69) | 0.36 |
| | Distal | —[a] | 17.06 (1.16) | —[a] |
| Posterior | Proximal | 42.25 (1.75) | 43.03 (1.99) | 0.22 |
| | Middle | 31.07 (1.78) | 30.95 (1.62) | 0.84 |
| | Distal | 18.90 (1.47) | 19.72 (1.13) | 0.86 |
| Lateral | Proximal | 23.92 (1.03) | 23.47 (0.97) | 0.17 |
| | Middle | 35.26 (0.97) | 34.94 (0.94) | 0.34 |
| | Distal | 53.99 (1.32) | 54.22 (1.03) | 0.55 |
| Medial | Proximal | 38.38 (2.19) | 38.29 (2.05) | 0.8 |
| | Middle | 12.67 (1.47) | 12.69 (0.87) | 0.97 |
| | Distal | 56.06 (2.44) | 56.16 (2.54) | 0.89 |

## 1.6 DISCUSSIONS

As shown in Figure 1.23, the FBG sensor liner produces more linear results than the F-Socket during the stance phase of gait, as shown by the smooth curves. Since F-Socket mats are resistive, their response is non-linear, even if it appears linear over a limited range (Brimacombe et al., 2009). The FBG-instrumented liner could be

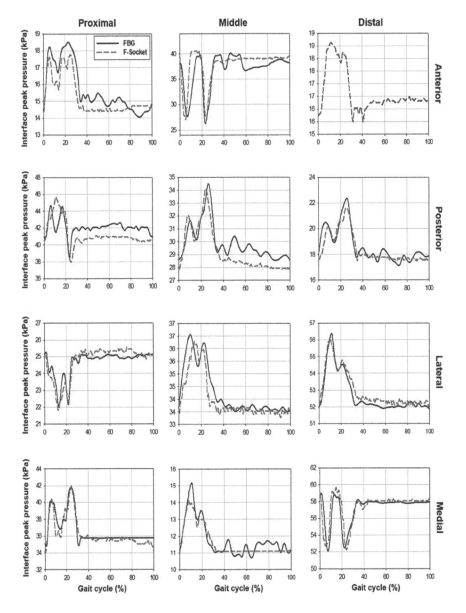

**FIGURE 1.23** The interface peak pressures at 12 subregions of the residual limb were measured using an FBG sensing liner and F-socket transducer during a full gait period.

used repeatedly during tests on amputee subjects without affecting measurement accuracy significantly.

The F-Socket mats have the advantage of allowing you to track and calculate pressures all over the residual limb, and each F-Socket mat is 20.3cm by 7.6cm in size. As a result, for amputees with a greater residual limb diameter, it may not be enough to cover the entire residual limb. As a result, the uses four F-Socket mats at the same time, which increases the cost. But, once mass manufactured in various sizes, FBG-instrumented liners can cover the entire residual limb and conform well to its surface, which can have an abnormal shape.

## 1.7  CONCLUSIONS

The FBG sensors exhibited very good reliability when tested *in situ* at the PT bar region. These promising results encouraged the author to further investigate the FBG sensing pad design which can exhibit the finest accomplishment when placed within the prosthetic socket.

According to the results, those fabrication parameters have an important impact on the overall performance of the FBG sensing pads. The FBG embedding depth, sensing pad thickness and embedding material were all taken into account in this analysis. The best sensing pad configuration for high sensitivity, excellent repeatability and low hysteresis error is to insert the FBGs in harder and thicker materials.

The results showed that four custom expandable sensing pads with at least two FBGs per pad successfully recorded the pressure data at eight subregions of the amputee's residual limb and could detect all the events of amputee's gait. These results were clinically validated by recording pressure simultaneously using the F-Socket mats. The pressure patterns recorded at all the subregions were similar for both sensor types. However, the FBG sensors reported higher pressure values than the F-socket.

The assessment of liner performance using a custom gait simulating machine showed that this liner could accurately map pressure distribution at the residual limb-socket interface. The statistical findings revealed that both mapping techniques could measure the pressure in a consistent and similar way (p-values $\leq 0.05$), implying that the FBG liner offers the same sensitivity and accuracy as the F-Socket mats. Furthermore, since this liner was designed and developed in the form of a prosthetic liner, the amputee will wear it comfortably during the experiments. Also, it can be used repeatedly for testing, without any significant effect on the measurement accuracy.

## REFERENCES

Abu Osman, N., Spence, W., Solomonidis, S., Paul, J., and Weir, A. (2010a). The patellar tendon bar! Is it a necessary feature? *Medical Engineering and Physics*, 32(7): 760–765.
Abu Osman, N., Spence, W., Solomonidis, S., Paul, J., and Weir, A. (2010b). Transducers for the determination of the pressure and shear stress distribution at the stump—socket interface of trans-tibial amputees. *Proceedings of the Institution of Mechanical Engineers, Part B: Journal of Engineering Manufacture*, 224(8): 1,239–1,250.

Al-Fakih, E., Abu Osman, A. A., Mahamd Adikan, F. R., Eshraghi, A., and Jahanshahi, P. (2016a). Development and validation of fiber Bragg grating sensing pad for interface pressure measurements within prosthetic sockets. *Sensors Journal, IEEE, 16*(4): 965–974.

Al-Fakih, E. A., Abu Osman, N. A., and Mahmad Adikan, F. R. (2016b). Techniques for interface stress measurements within prosthetic sockets of transtibial amputees: a review of the past 50 years of research. *Sensors, 16*(7): 1,119.

Ali, S., Abu Osman, N. A., Mortaza, N., Eshraghi, A., Gholizadeh, H., and Abas, W. A. B. B. W. (2012). Clinical investigation of the interface pressure in the trans-tibial socket with Dermo and Seal-In X5 liner during walking and their effect on patient satisfaction. *Clinical Biomechanics, 27*(9): 943–948.

Baars, E., and Geertzen, J. (2005). Literature review of the possible advantages of silicon liner socket use in trans-tibial prostheses. *Prosthetics and Orthotics International, 29*(1): 27–37.

Brimacombe, J. M., Wilson, D. R., Hodgson, A. J., Ho, K. C., and Anglin, C. (2009). Effect of calibration method on Tekscan sensor accuracy. *Journal of Biomechanical Engineering, 131*(3): 034503.

Dou, P., Jia, X., Suo, S., Wang, R., & Zhang, M. (2006). Pressure distribution at the stump/ socket interface in transtibial amputees during walking on stairs, slope and non-flat road. *Clinical Biomechanics, 21*(10): 1067–1073.

Dumbleton, T., Buis, A. W., McFadyen, A., McHugh, B. F., McKay, G., Murray, K. D., & Sexton, S. (2009). Dynamic interface pressure distributions of two transtibial prosthetic socket concepts. *Journal of Rehabilitation Research & Development, 46*(3).

Eshraghi, A., Osman, N. A. A., Gholizadeh, H., Ali, S., Sævarsson, S. K., and Abas, W. A. B. W. (2013). An experimental study of the interface pressure profile during level walking of a new suspension system for lower limb amputees. *Clinical Biomechanics, 28*(1): 55–60.

Eshraghi, A., Osman, N. A. A., Gholizadeh, H., Karimi, M., & Ali, S. (2012). Pistoning assessment in lower limb prosthetic sockets. *Prosthetics and Orthotics International, 36*(1), 15–24.

Friel, K. (2005). Componentry for lower extremity prostheses. *JAAOS-Journal of the American Academy of Orthopaedic Surgeons, 13*(5): 326–335.

Galdik, J. (1955). The below knee suction socket. *Prosthetic Appliance Journal, 9*: 43–46.

Laing, S., Lee, P. V., and Goh, J. C. (2011). Engineering a trans-tibial prosthetic socket for the lower limb amputee. *Annals of the Academy of Medicine, Singapore, 40*(5): 252–259.

Letchuman, G., Wan Nazaimoon, W., Wan Mohamad, W., Chandran, L., Tee, G., Jamaiyah, H., Zanariah, Isa, Fatanah, H., and Ahmad Faudzi, Y. (2010). Prevalence of diabetes in the malaysian national health morbidity survey III 2006. *Medical Journal of Malaysia, 65*(3): 180–186.

Luyckx, G., Voet, E., De Waele, W., & Degrieck, J. (2010). Multi-axial strain transfer from laminated CFRP composites to embedded Bragg sensor: I. Parametric study. *Smart Materials and Structures, 19*(10): 105017.

Moo, E. K., Abu Osman, N. A., Pingguan-Murphy, B., Wan Abas, W. A. B., Spence, W. D., & Solomonidis, S. E. (2009). Interface pressure profile analysis for patellar tendon-bearing socket and hydrostatic socket. *Acta of Bioengineering and Biomechanics, 11*(4): 37–43.

Ng, P., Lui, L., Lee, V. S. P., Tan, K. C., Tay, E. H., & Goh, J. C. (2002). *Rapid manufacturing machine (RMM) for prosthetic socket fabrication.* In *Proceedings of the 10th International Conference on Biomedical Engineering* (pp. 417–418).

NHMS. (2015). The national health and morbidity survey, Ministry of Health-Malaysia. Retrieved on March 27 2016 from http://www.iku.gov.my/index.php/research-eng/list-of-research-eng/iku-eng/nhms-eng/nhms-2015.

Pirouzi, G., Abu Osman, N. A., Eshraghi, A., Ali, S., Gholizadeh, H., & Wan Abas, W. A. B. (2014). Review of the socket design and interface pressure measurement for transtibial prosthesis. *The Scientific World Journal, 2014*: 849073.

Polliack, A. A., Sieh, R. C., Craig, D. D., Landsberger, S., McNeil, D. R., & Ayyappa, E. (2000). Scientific validation of two commercial pressure sensor systems for prosthetic socket fit. *Prosthetics and Orthotics International, 24*(1): 63–73.

Radcliffe, C. W. (1962). The biomechanics of below-knee prostheses in normal, level, bipedal walking. *Artificial Limbs, 6*, 16–24.

Radcliffe, C. W. & Foort J. (1961). *The patellar-tendon-bearing below-knee prosthesis.* Biomechanics Laboratory, University of California, Berkeley and San Francisco.

Reiber, G. E., (1994). Who is at risk for limb loss and what to do about it?.. *Journal of Rehabilitation Research and Development, 31:* 357–357.

Safari, M. R., and Meier, M. R. (2015). Systematic review of effects of current transtibial prosthetic socket designs—Part 1: Qualitative outcomes. *Journal of Rehabilitation Research and Development, 52*(5): 491–510.

Sengeh, D. M., & Herr, H. (2013). A variable-impedance prosthetic socket for a transtibial amputee designed from magnetic resonance imaging data. *JPO: Journal of Prosthetics and Orthotics, 25*(3): 129–137.

Wu, C. L., Chang, C. H., Hsu, A. T., Lin, C. C., Chen, S. I., & Chang, G. L. (2003). A proposal for the pre-evaluation protocol of below-knee socket design-integration pain tolerance with finite element analysis. *Journal of the Chinese Institute of Engineers, 26*(6), 853–860.

Zhang, M., Turner-Smith, A., Tanner, A., and Roberts, V. (1998). Clinical investigation of the pressure and shear stress on the trans-tibial stump with a prosthesis. *Medical Engineering and Physics, 20*(3): 188–198.

# 2 Magnetic Suspension System (MPSS)

## N A Abu Osman

University of Malaya, Kuala Lumpur, Malaysia

## A Eshraghi

Prosthetic and Orthotic Services, West Park Healthcare Centre, Toronto, Canada

## CONTENTS

## 2.1 BACKGROUND

The main aim of recovery for lower limb amputees is to re-establish as much natural gait as possible. The most suitable components should be used in prosthetic devices to enable normal gait work. Momentum, gravity and other ambulation forces appear to displace the prosthesis on the residual limb, most notably during the swing process of the gait (Smith et al., 2004). As a result, various systems for securely suspending the prosthetic leg on the residual limb have been established. The proper suspension system will boost the amputee's gait and reduce energy expenditure (Schmalz et al., 2002). Bad suspension has many effects, including gait deviation, vertical movement or pistoning inside the socket, pain, skin breakdown and reduced user satisfaction (Grevsten, 1978; Narita et al., 1997; Kapp, 1999; Dillingham et al., 2001; Schmalz et al., 2002, Bruno and Kirby, 2009).

Suspension is achieved anatomically or externally through various components. Supracondylar/suprapatellar system, supracondylar system, supracondylar system, or PTS, suprapatellar strap, thigh corset, waist belt, sleeve, suction, or vacuum and locking liner are examples of these systems. Suspension is accomplished in the case of osseointegration by directly attaching the prosthesis to the residual bone (Webster et al., 2009).

DOI: 10.1201/9781003196730-2

The suspension mechanism is designed to prevent the residual limb from translating, rotating, and pistoning (vertical movement) in relation to the prosthesis socket. Previous research has found that inadequate suspension has a negative impact on recovery, as well as the comfort and activity level of lower limb amputees (Kristinsson, 1993; Van de Weg and Van der Windt, 2005). According to a study of 146 prosthetic users, the majority were dissatisfied with their prostheses due to skin problems and discomfort (Dillingham et al., 2001). Kark and Simmons (2011) discovered that amputee participants were dissatisfied with their prostheses in their research (Kark and Simmons, 2011). Overall, 77% of users were dissatisfied with the polyethylene foam liner compared to the pin/lock system (Coleman et al., 2004). A prospective research, on the other hand, found that almost all the participants preferred the polyethylene foam liner (Boonstra et al., 1996). To explain these contentious findings, the research team, which included the author of this thesis, conducted a retrospective analysis. The majority of lower limb amputees were found to be disappointed with their prostheses. A questionnaire survey was administered to 243 males who had undergone traumatic unilateral transtibial amputation. The pin/lock system, polyethylene foam liner and Seal-In suspension were listed as the most widely used suspension systems. Suspension device satisfaction is a multifaceted problem. The updated prosthetic assessment questionnaire (PEQ) surveyed common problems and satisfaction items based on the literature.

The results revealed that the Seal-In suspension was preferred by the participants over the pin/lock and the polyethylene foam liner (Gholizadeh et al., 2013). Except for sweating, there were significant differences in the perceived problems among the suspension systems; primarily, the polyethylene foam and the Seal-In suspension caused excessive sweating. The Seal-In device received higher overall satisfaction ratings than the pin/lock and polyethylene foam liner. Participants also favoured the pin/lock and Seal-In liner over the polyethylene liner, which refuted Boonstra's et al. (1996) and Coleman et al. (2004). Both researchers found that the pin/lock system was ineffective; though, McCurdie et al. (1997) The preference for the pin/lock has been clearly defined (McCurdie et al., 1997). Later, Van der Linde et al. (2004) professionals in the field of prosthetics preferred the pin/lock scheme, as well (Van der Linde et al., 2004).

Åström and Stenström (2004) and Hatfield and Morrison (2001) revealed that amputees preferred the pin/lock liner over the polyethylene foam liner (Åström and Stenström, 2004; Hatfield and Morrison, 2001). In our research, participants were more comfortable with the pin/lock and the Seal-In suspension when walking (even and rough ground) and negotiating stairs (Gholizadeh et al., 2013). Improved cosmesis and suspension of the prosthesis improve function and satisfaction (Wirta et al., 1990). In our research, the pin/lock and Seal-In suspensions outperformed the polyethylene foam liner in terms of suspension. This outcome was coherent with the results of Cluitmans et al. (1994) and Baars and Geertzen (2005), who studied enhanced suspension with the pin/lock scheme (Baars and Geertzen, 2005; Cluitmans et al., 1994).

During the swing process of the gait, the pin/lock mechanism compresses the residual limb proximally and tensions it distally, according to research. Milking refers to the skin stretch at the pin spot. This milking phenomenon is most likely responsible for the observed short-term (oedema and redness) and long-term (discoloration and thickening) transformations, particularly at the distal end of the residuum (Beil and Street, 2004). Pressure, discomfort and residual limb atrophy or volume loss may result from this compression. Likewise, the satisfaction survey contrasted the Seal-In and pin/lock systems and discovered that the pin/lock system caused more pain (Gholizadeh et al., 2013).

Previous research identified the key issues of lower limb amputees with common suspension systems using soft silicone liners. The survey studies conducted by the author of this study also revealed the debate about the best suspension device for lower limb prostheses. There was no single suspension system that was shown to be effective in any way for all types of users. For example, one device was easier to put on and take off but painful to walk in, while the other caused the opposite problem. The existing suspension systems, which include mechanical parts, have a low reliability and need a lot of maintenance (Gholizadeh et al., 2013). Based on the findings of the literature reviews and survey studies, it is determined that developing a new suspension system is needed to improve the positive aspects of the current systems while minimizing their disadvantages. As a result, this study concentrated on the design, development and *in vivo* testing of a suspension system for lower limb amputees.

## 2.2   MECHANICAL MAGNETIC SUSPENSION SYSTEM

Prosthetic sockets are essential components of prosthetics. Suspension systems are essentially connecting devices located between the prosthetic socket and the distal components of the prosthesis, such as the prosthetic foot, ankle and pylon. Durability, cosmetic appearance, comfort, function and cost are the most important considerations to consider when designing prostheses. With these considerations in mind, the concept was completed using a 3D mechanical computer-aided design (CAD) software (SolidWorks, 2009). Any lower limb prosthesis for people who have had their lower limbs amputated includes, but is not limited to, the socket, pylon (shank), knee, and foot (Kutz et al., 2003). The prosthetic suspension device is usually installed either inside the socket or between the socket and the pylon. Given the limited space available at this interface, the MPSS used in this study was designed to suit the socket end of an adult amputee (Figure 2.1). The coupling system's height was also restricted by the available space, allowing it to be used with long residual limbs. Since silicone liners are readily available and widely used, the MPSS was designed to be used with them (Eshraghi et al., 2013a). To that end, a cap was created that suited both the main body of the new coupling system and the distal end of the liner. The dimensions were purposefully designed to fit those of the liner. The cross section was square, and the cap was hollow to save weight. The hollow space featured a central screw in the centre and was filled with silicone adhesive to ensure a secure connection to the liner. The magnetic field served as the foundation for the coupling concept. As a result, the cap was made of mild steel to have a strong grip. The magnetic power was generated by the coupling device's body. A permanent Neodymium Iron Boron magnet was used, which was small but had a high magnetic strength.

The magnetic field was amplified by the housing's flanges. A mechanical switch was attached to the housing and the magnet to control the magnetic strength. When

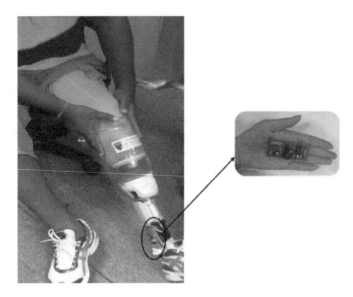

**FIGURE 2.1** Magnetic Suspension System (MPSS). A participant is donning a prosthesis that is equipped with the MPSS and the coupling alarm.

the rotary switch was in the "on" position, the cap was attracted to the casing, while when the switch was in the "off" position, it was removed from the lower body of the coupling system.

A coupling warning system capable of detecting any malfunction in the coupling system for the lower limb prosthesis was developed (Eshraghi et al., 2013a). An interface, a process unit and a power supply were all part of this device. The signals were detected and processed by a microcontroller device, which then decides whether or not to power on the output. The interface consisted of two inputs and one output. The magnetic field was detected by one Hall effect sensor, and a touch sensor ensured that the joint remained in complete contact with the leg. The output is a buzzer that is activated by a transistor to amplify the microcontroller signals and generate the necessary warning. The buzzer produced an audible warning signal at 97dB and a frequency of 2KHz. The microcontroller needed power in the 2.5–5V range.

A 9V battery was needed for the buzzer and Hall effect sensor. As a result, a 9V battery, a 1V regulator and two transistors were used. The transistors switched the voltage between the microprocessor output and the desired voltage for the Hall effect sensor and buzzer. The microprocessor was necessary to determine whether the coupling was effective or not. If the Hall effect sensor signals indicate that the magnetic field has triggered the coupling, the touch sensor signals are examined.

For 3 milliseconds, the microcontroller sampled each signal millisecond by millisecond. If all the data are the same, it will be replaced by the previous (Figure 2.2). This procedure was repeated three times to ensure that the sensor sensed the vibration of the coupling rather than the detachment. The final result was analyzed by the microprocessor so that an acceptable decision could be made. This unit came with a single 1,200mAh, 9V battery.

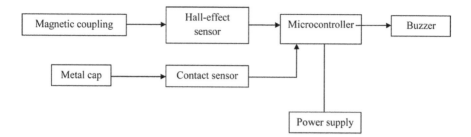

**FIGURE 2.2** Alarm system. The coupling warning system is depicted as a block diagram.

## 2.3 EXPERIMENTAL FINDINGS

During ambulation, the MPSS could safely suspend the prosthesis on the amputee's residual limb (Eshraghi et al., 2013a). For the first time, a fire warning device was used in a prosthetic suspension system. The findings of this study demonstrated that the magnetic coupling system could withstand the forces that appear to displace the prosthesis on the residual limb when walking. Any failure of the prosthesis coupling can result in serious trauma, exacerbating amputees' difficulties. To avoid this, the prosthesis suspension system included an optional warning system. To the best of the author's knowledge, this is the first suspension device for prosthesis that includes a safety system. This study's suspension system can be used with or without the alarm system.

In general, pistoning with the MPSS was within the ranges seen for the other suspension systems tested, and the values showed some improvement on some gait phases (Eshraghi et al., 2012a). The results of this study show that the form of suspension system has a major effect on pistoning during gait. The rates of pistoning for the pin lock system and the magnetic suspension system were similar during the gait simulation and in actual walking, but higher in contrast to the suction system. Nonetheless, all the systems demonstrated pistoning values that were within the ranges recorded in the literature. The measurements were highly repeatable, with no noticeable differences between observers. The Seal-In suspension system was the most effective in reducing vertical movement during level walking (Eshraghi et al., 2012b).

The MPSS can reduce pressure on the residual limb, particularly during the swing phase of gait. In a prosthetic socket, an even distribution of pressure is thought to be optimal. During the swing process, users of the pin/lock liners feel a stretch at the distal tissue of the residual limb (milking phenomenon). When compared to the pin/lock, the MPSS developed lower peak pressures at the anterior and posterior surfaces during the swing phase of gait (Table 2.1) (Eshraghi et al., 2013b). During level walking, stair and slope negotiation, the Seal-In suspension produced significantly higher peak pressures than the pin/lock and MPSS systems (Figure 2.3) (Eshraghi et al., 2015).

As opposed to the MPSS and pin/lock, the Seal-In liner focused pressure primarily at the middle and distal regions of the posterior sensor. This may be because of the position of the seals and the fact that suction is produced primarily at the distal end, where the valve is located. During the stair and ramp agreement, the intra-system pressure distribution at the anterior and posterior surfaces of the residual limb was somewhat homogeneous for all suspension systems. Despite this, inter-system

**TABLE 2.1**

**Average Peak Pressure (kPa) at Anterior, Posterior, Medial and Lateral Residual Limb Sensor Sites during Level Walking**

| Suspension Type | Ant Mean (SD) | Pos Mean (SD) | Med Mean (SD) | Lat Mean (SD) |
|---|---|---|---|---|
| Pin/lock[1] | 89.89 | 47.22 | 39.21 | 31.65 |
| | (26.4) | (17.7) | (18.1) | (15.2) |
| New magnetic lock[2] | 79.26 | 26.01 | 38.07 | 27.41 |
| | (23.2) | (13.3) | (12.5) | (9.8) |
| Seal-in liner[3] | 119.43 | 65.29 | 53.50 | 52.55 |
| | (30.8) | (16.6) | (21.7) | (14.5) |
| Sig. (two tailed)* | 1–2 (0.042) | 1–2 (0.003) | 1–3 (0.034) | 1–3 (0.023) |
| | 1–3 (0.017) | 1–3 (0.011) | 2–3 (0.027) | 2–3 (0.015) |
| | 2–3 (0.026) | 2–3 (0.000) | | |

Ant = Anterior; Pos = Posterior; Med = Medial; Lat = Lateral.

* The numbers "1–2", "1–3" and "2–3" suggest that major variations (P0.05) were discovered between the two suspension systems using paired-samples t-tests.

pressure mapping revealed significant differences between the suspension types, particularly at the anterior and posterior sensor sites (Figures 2.4 and 2.5).

In contrast to the Seal-In and pin/lock systems, the MPSS demonstrated superior qualities in certain gait kinetic and kinematic parameters (Eshraghi et al., 2014). Some gait kinetics and kinematics were altered by the suspension type; major variations were observed in the GRF (vertical and fore-aft), knee and ankle angles (Table 2.2). The type of suspension had no discernible effect on walking pace, stance or swing time. According to the GRF results, the MPSS can reduce loading over the proximal limb joints when compared to the pin/lock system.

The GDI scores showed that the amputees' gait kinematics were inferior to that of normal individuals; however, the three suspension systems demonstrated comparable clinical results that enabled the amputees to ambulate successfully (Figure 2.6). This demonstrates that the MPSS is a comparable suspension choice for lower limb amputees.

On a variety of clinically important products, the MPSS resulted in higher satisfaction scores than the pin/lock and Seal-In schemes (Table 2.3). All the suspension systems investigated in this study received relatively high levels of satisfaction from participants (Eshraghi et al., 2012a). Nonetheless, the qualitative survey found substantial variations in satisfaction and perceived issues with the MPSS as opposed to the pin lock and suction systems. The MPSS made less noise when walking and donning than the pin/lock suspension, was much easier to don and doff than the Seal-In suspension and pin/lock system and resulted in higher overall satisfaction as compared to the Seal-In and pin/lock systems.

"1–2", "1–3" and "2–3" indicate that significant differences ($P < 0.05$) were found between each two suspension systems in each satisfaction/problems item based on the paired-samples t tests.

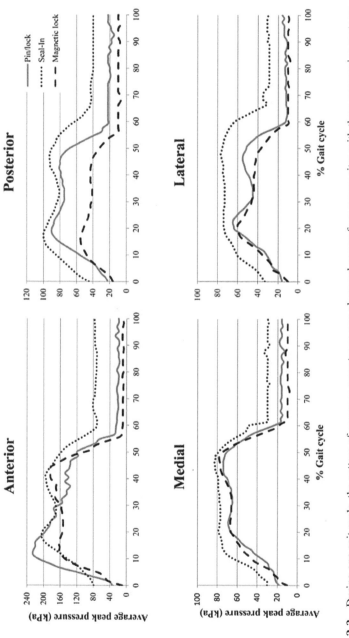

**FIGURE 2.3** During one gait cycle, the pattern of pressure acceptance was observed across four sensor sites with three suspension systems.

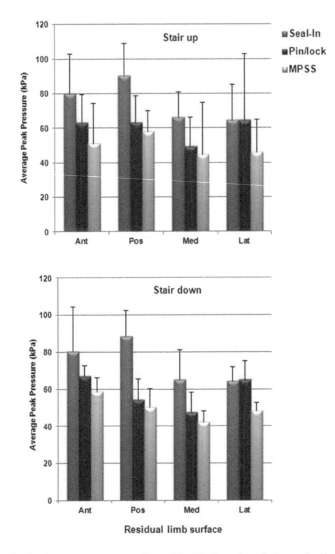

**FIGURE 2.4**  Peak pressure pattern at the residual limb surface during stair climbing.

When opposed to the pin/lock suspension, users felt more comfortable with the MPSS. The MPSS's cosmesis was almost identical to that of the pin/lock system. In terms of cosmesis, the subjects were more pleased with the suction system than with the MPSS, which can be due to the additional components used for the magnetic suspension. The participants in this study were mostly unhappy with the Seal-In system's donning and doffing; donning and doffing was substantially easier with the magnetic system. For long-term usage, participants preferred the MPSS over the Seal-In and pin/lock. As compared to the pin lock suspension, the MPSS caused substantially less discomfort in the participants (Eshraghi et al., 2012a).

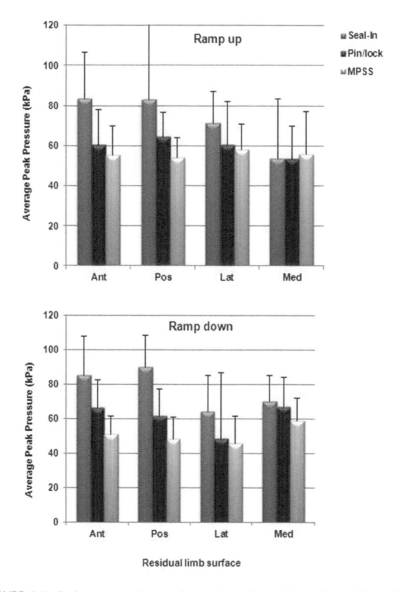

**FIGURE 2.5** Peak pressure values at four main residual limb surfaces while walking downhill.

In short, as opposed to the pin/lock and MPSS systems, the Seal-In system resulted in less pistoning. With the Seal-In interface device, there could be a relationship between higher peak pressure and low pistoning. This study found higher pressure magnitudes with the Seal-In method, which may explain lower amounts of pistoning previously observed. While suction devices, such as the Seal-In, may improve prosthetic fit, the improved fit and resulting increased pressure may cause residual limb atrophy, skin problems and interruptions in blood flow to the limb. It is possible to

**TABLE 2.2**

**Differences in Kinetic and Kinematic Properties between Sound and Prosthetic Limbs within Each Suspension Form; Mean (95% CI)**

| Parameters | Seal-In | | | | | Pin/lock | | | | | MPSS | | | | |
|---|---|---|---|---|---|---|---|---|---|---|---|---|---|---|---|
| | Sound | Prosthesis | P value | MD (CI) | d | Sound | Prosthesis | P value | MD (CI) | d | Sound | Prosthesis | P value | MD (CI) | d |
| Step length (m) | 0.57 (0.53 –0.61) | 0.61 (0.55 –0.66) | 0.320 | 0.04 (−0.84 –3.09) | 0.1 | 0.54 (0.47 –0.62) | 0.62 (0.54 –0.69) | 0.134 | 0.08 (−0.03 –0.17) | 0.5 | 0.56 (0.5 –0.62) | 0.59 (0.51 –0.67) | 0.536 | 0.03 (−0.07 –0.12) | 0.2 |
| Cadence (step/min) | 94.09 (92.73 –95.46) | 95.21 (94.02 –96.41) | 0.183 | 1.12 (−0.85 –3.09) | 0.2 | 93.03 (91.77 –94.3) | 95.60 (94.13 –97.25) | **0.031** | 2.57 (0.28 –4.03) | 0.4 | 93.03 (91.77 –94.3) | 95.06 (93.37 –96.75) | 0.145 | 2.03 (−0.73 –4.4) | 0.6 |
| Stance time (% of gait cycle) | 65.56 (64.1 –67.03) | 62.28 (60.89 –63.70) | **0.002** | 3.28 (−5.11 –−1.45) | 1.3 | 66.7 (65.53 –67.87) | 60.73 (59.74 –61.73) | **<0.001** | 5.97 (−7.38 –−4.55) | 3.4 | 65.57 (64.34 –66.8) | 62.31 (61.19 –63.42) | **0.001** | 3.26 (−4.77 –−1.75) | 1.7 |
| Swing time (% of gait cycle) | 34.46 (33.31 –35.61) | 37.70 (65.60 –67.80) | **0.001** | 32.24 (30.74 –33.75) | 1.4 | 33.32 (31.64 –35) | 38.30 (36.95 –39.65) | **<0.001** | 4.98 (3.32 –6.64) | 2.1 | 34.14 (32.75 –35.52) | 37.56 (36.39 –38.73) | **0.001** | 3.42 (1.83 –5.02) | 1.7 |
| Vertical GRF, 1st peak (%BW) | 121.11 (118.05 –124.17) | 99.68 (97.15 –102.22) | **<0.001** | 21.43 (−25.15 –−17.7) | 4.8 | 126.68 (123.88 –129.48) | 104.22 (101.58 –106.87) | **<0.001** | 22.46 (−26.03 –−18.89) | 4.9 | 115.27 (109.13 –121.42) | 96.42 (91.84 –101.02) | **<0.001** | 18.85 (−25.2 –−12.49) | 2.3 |
| Vertical GRF, 2nd peak (%BW) | 101.99 (99.59 –104.4) | 102.63 (100.19 –105.06) | 0.706 | 0.64 (−1.69 –2.96) | 0.1 | 101.12 (98.87 –103.38) | 99.09 (96.34 –101.85) | 0.301 | 2.03 (−6.12 –2.06) | 0.4 | 105.18 (102.38 –107.98) | 91.69 (88.51 –94.87) | **<0.001** | 13.49 (−17.49 –−9.49) | 2.4 |
| Fore-aft GRF, 1st peak (%BW) | 7.86 (7.1 –8.62) | 5.45 (4.79 –6.12) | **<0.001** | 2.41 (−3.34 –−1.47) | 2.1 | 9.34 (8.4 –10.28) | 4.66 (3.98 –5.35) | **<0.001** | 4.68 (−5.7 –−3.66) | 3.9 | 9.86 (8.94 –10.78) | 4.11 (3.43 –4.80) | **<0.001** | 5.75 (−6.87 –−4.61) | 4.8 |
| Fore-aft GRF, 2nd peak (%BW) | −7.51 (−8.25 –−6.77) | −8.10 (−8.76 –−7.43) | 0.208 | 0.59 (−1.37 –0.19) | 0.5 | −7.13 (−8.84 –−6.45) | −8.11 (−8.91 –−7.31) | 0.058 | 0.98 (−2.04 –0.04) | 0.7 | −7.01 (−8.10 –−6.25) | −7.41 (−8.13 –−6.69) | 0.390 | 0.40 (−1.25 –−0.52) | 0.3 |

| | | | | | | | | | | | | | | | |
|---|---|---|---|---|---|---|---|---|---|---|---|---|---|---|---|
| Hip position-initial contact | 35.89 (33.81, −37.97) | 32.8 (30.95, −34.65) | 0.193 | 3.09 (−5.38, −−0.8) | 0.9 | 32.6 (30.94, −34.26) | 33.11 (31.04, −35.17) | 0.543 | 0.51 (−1.26, −2.27) | 0.2 | 34.15 (32.11, −35.81) | 33.04 (31.08, −35.00) | 0.318 | 1.11 (−3.44, −1.21) | 0.4 |
| Max Hip Ext | −2.13 (−2.46, −−1.81) | 3.06 (2.71, −3.42) | **<0.001** | 5.19 (4.83, −5.56) | 3.6 | −2.42 (−2.98, −−1.85) | 2.62 (2.18, −3.05) | **<0.001** | 5.04 (4.36, −5.71) | 3.6 | −2.42 (−2.75, −−1.67) | 2.5 (1.97, −3.04) | **<0.001** | 4.92 (4.29, −5.53) | 5.4 |
| Hip ROM | 38.42 (37.37, −39.47) | 37.31 (35.83, −38.79) | 0.193 | 1.11 (−2.66, −0.43) | 0.5 | 37.23 (35.03, −38.80) | 36.13 (34.92, −37.33) | 0.121 | 1.1 (−2.55, −0.34) | 0.5 | 37.52 (35.67, −39.45) | 36.7 (35.25, −38.16) | 0.261 | 0.82 (−2.18, −0.65) | 0.4 |
| Knee position-initial contact | 1.41 (1.14, −1.67) | 5.4 (4.55, −6.25) | **<0.001** | 3.99 (3.12, −4.87) | 3.8 | 4.1 (3.17, −5.02) | 5.73 (4.9, −6.57) | **0.022** | 1.63 (0.28, −2.99) | 1.1 | 3.9 (3.35, −4.45) | 5.53 (4.34, −6.71) | **0.023** | 1.63 (0.27, −2.98) | 1.1 |
| Max Knee Flex – stance | 15.12 (14.09, −16.15) | 13.72 (12.59, −14.86) | 0.059 | 1.40 (−2.98, −0.18) | 0.8 | 13.43 (11.86, −15.01) | 12.47 (11.08, −13.85) | 0.302 | 0.96 (−2.93, −0.99) | 0.4 | 14.24 (12.66, −15.82) | 12.84 (11.5, −14.19) | 0.235 | 1.40 (−3.83, −1.04) | 0.6 |
| Max Knee Flex- swing | 55.17 (53.58, −56.75) | 75.40 (73.21, −77.57) | **<0.001** | 20.23 (17.32, −23.13) | 6.4 | 52.52 (51.08, −53.96) | 66.92 (64.77, −69.08) | **<0.001** | 14.4 (11.49, −17.32) | 4.7 | 54.02 (52.06, −55.97) | 70.81 (68.7, −72.93) | **<0.001** | 16.79 (14.21, −19.38) | 5.0 |
| Knee ROM | 56.14 (54.57, −57.7) | 70.68 (68.34, −73.04) | **<0.001** | 14.54 (11.54, −17.57) | 4.4 | 52.61 (51.12, −54.09) | 61.42 (58.99, −63.81) | **<0.001** | 8.81 (6.28, −11.31) | 2.7 | 52.79 (51.28, −54.3) | 58.25 (56.55, −59.94) | **<0.001** | 5.46 (3.02, −7.89) | 2.1 |
| Ankle position-initial contact | 2.12 (1.59, −2.65) | −0.81 (−1.21, −−0.41) | 0.583 | 2.93 (−3.67, −−2.19) | 3.8 | −4.21 (−4.88, −−3.54) | 0.27 (0.07, −0.46) | **<0.001** | 4.48 (3.76, −5.19) | 5.5 | −2.29 (−2.81, −−1.77) | −0.6 (−0.93, −−0.28) | **<0.001** | 1.69 (1.01, −2.37) | 2.3 |
| Max ankle PF-stance | −6.68 (−8.33, −−5.02) | −7.19 (−8.3, −−6.07) | 0.583 | 0.51 (−2.75, −1.73) | 0.2 | −5.92 (−7.23, −−4.62) | −5.89 (−6.98, −−4.81) | 0.951 | 0.03 (−1.09, −1.15) | 0.0 | −6.12 (−7.41, −−4.82) | −3.02 (−3.73, −−2.31) | **0.002** | 3.10 (1.42, −4.77) | 1.8 |

(Continued)

**TABLE 2.2** (*Continued*)

**Differences in Kinetic and Kinematic Properties Between Sound and Prosthetic Limbs within each Suspension Form; Mean (95% CI)**

| Parameters | Seal-In | | | | | Pin/lock | | | | | MPSS | | | | |
|---|---|---|---|---|---|---|---|---|---|---|---|---|---|---|---|
| | Sound | Prosthesis | P value | MD (CI) | d | Sound | Prosthesis | P value | MD (CI) | d | Sound | Prosthesis | P value | MD (CI) | d |
| Max ankle DF-stance | 7.3 (6.23 – 8.37) | 14.49 (13.34 – 15.63) | **<0.001** | 7.19 (5.44 – 8.93) | 3.9 | 8.09 (7.07 – 9.1) | 15.11 (14.24 – 15.98) | **<0.001** | 7.02 (5.72 – 8.32) | 4.5 | 7.92 (6.78 – 9.06) | 14.67 (13.93 – 15.41) | **<0.001** | 6.75 (5.43 – 8.06) | 4.2 |
| Max ankle PF-swing | -13.2 (-14.7 – -11.7) | 0.33 (0.12 – -0.55) | **<0.001** | 13.53 (12.02 – 15.05) | 7.6 | -12.15 (-13.2 – -11.1) | 1.37 (1.13 – -1.67) | **<0.001** | 13.52 (12.45 – 14.64) | 5.7 | -12.17 (-13.05 – -11.29) | 1.13 (0.93 – -1.33) | **<0.001** | 13.30 (12.38 – 14.21) | 5.2 |
| Ankle ROM | 20.67 (19.1 – 22.24) | 21.73 (20.35 – 23.1) | 0.280 | 1.06 (-1.23 – -3.35) | 0.4 | 20.08 (18.68 – 21.48) | 20.87 (19.32 – 22.43) | 0.508 | 0.79 (-1.74 – -3.33) | 0.3 | 20.25 (18.5 – 21.99) | 20.69 (19.55 – 21.83) | 0.700 | 0.44 (-2.00 – -2.88) | 0.2 |

CI = confidence interval; PF = plantar flexion; DF = dorsiflexion; Flex = flexion; Ext = extension; ROM = range of motion; MD = mean difference.

Values of significance ($P < 0.05$) have been shown in bold. $d$ equates to values of Cohen's d: 0.2 = small, 0.5 = medium, >0.8 = large.

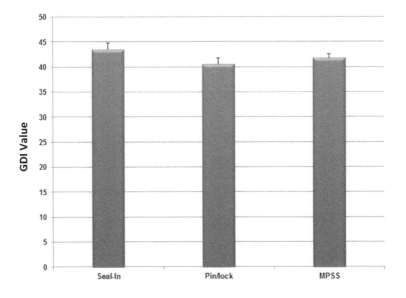

**FIGURE 2.6**  A comparison of GDI values between suspension systems. The error bars reflect the normal error values.

## TABLE 2.3
## Satisfaction and Issue Ratings for Three Suspension Systems

| | | Suction (1) | Magnetic lock (2) | Pin/lock (3) | Sig. (t test) |
|---|---|---|---|---|---|
| **Satisfaction** | Fitting | 87.09 | 76.82 | 79.59 | 1–2(0.002) 1–3(0.003) 2–3(0.044) |
| | Donning and doffing | 57.24 | 79.68 | 71.44 | 1–2(0.000) 1–3(0.016) 2–3(0.000) |
| | Sitting | 79.41 | 76.44 | 68.80 | 1–3(0.001) 2–3(0.041) |
| | Walking | 65.21 | 84.66 | 72.80 | 1–2(0.000) 1–3(0.008) 2–3(0.000) |
| | Uneven walking | 63.91 | 77.93 | 54.30 | 1–2(0.000) 1–3(0.030) 2–3(0.000) |
| | Stair | 68.83 | 80.60 | 65.75 | 1–2(0.001) 2–3(0.000) |
| | Suspension | 93.71 | 81.72 | 75.20 | 1–2(0.000) 1–3(0.000) |
| | Cosmesis | 83.10 | 73.27 | 69.05 | 1–2(0.004) 1–3(0.000) |
| | Overall satisfaction | 63.14 | 83.10 | 75.94 | 1–2(0.000) 1–3(0.000) 2–3(0.008) |

*(Continued)*

**TABLE 2.3    (*Continued*)**
**Satisfaction and Issue Ratings for Three Suspension Systems**

|  |  | Suction (1) | Magnetic lock (2) | Pin/lock (3) | Sig. (t test) |
|---|---|---|---|---|---|
| **Problems** | Sweat | 64.78 | 60.16 | 55.00 | 1–2(0.009) 1–3(0.019) |
|  | Wound | 95.17 | 75.04 | 81.85 | 1–2(0.000) 1–3(0.000) 2–3(0.017) |
|  | Irritation | 94.66 | 75.10 | 81.28 | 1–2(0.000) 1–3(0.000) 2–3(0.021) |
|  | Pistoning within the socket | 96.47 | 63.95 | 84.18 | 1–2(0.000) 1–3(0.000) 2–3(0.000) |
|  | Rotation within the socket | 99.57 | 81.65 | 80.18 | 1–2(0.000) 1–3(0.000) 2–3(0.017) |
|  | Swelling | 94.91 | 89.64 | 86.75 | 1–2(0.017) 1–3(0.000) |
|  | Bad smell | 77.83 | 63.94 | 72.49 | 1–2(0.000) 1–3(0.007) 2–3(0.002) |
|  | Unwanted sound | 96.81 | 80.28 | 70.21 | 1–2(0.000) 1–3(0.000) 2–3(0.003) |
|  | Pain | 80.67 | 90.18 | 70.62 | 1–2(0.000) 1–3(0.000) 2–3(0.000) |

assume that pistoning alone is not a reliable predictor of clinically superior suspension systems. When selecting a prosthetic suspension system for a lower limb amputee, satisfaction, especially with donning and doffing, should be considered.

## 2.4   CONCLUSIONS

To summarize, since the needs and abilities of amputees differ greatly, the creation of new suspension systems will provide clinicians with more options for selecting the device that better meets their patients' needs. Biomechanical and clinical assessments were used to validate the benefits of our method. There are many low-cost components available. However, various experts do not agree on the key criterion for selecting a suitable suspension device that corresponds to the needs and abilities of amputees. Usually, prosthetic components are recommended based on practical experience. As a result, a dependable and objective criterion for prosthetic prescriptions should be established. This study's knowledge and experience, as well as methodical evaluation, will help an amputee choose a suitable form of prosthesis.

## REFERENCES

Åström, I., & Stenström, A. (2004). Effect on gait and socket comfort in unilateral transtibial amputees after exchange to a polyurethane concept. *Prosthetics and Orthotics International*, 28(1), 28–36.

Baars, E., & Geertzen, J. (2005). Literature review of the possible advantages of silicon liner socket use in transtibial prostheses. *Prosthetics and Orthotics International*, 29(1), 27–37.

Beil, T. L., & Street, G. M. (2004). Comparison of interface pressures with pin and suction suspension systems. *Journal of Rehabilitation Research & Development*, 41(6A), 821–828.

Boonstra, A., Van Duin, W., & Eisma, W. (1996). Silicone suction socket (3S) versus supra-condylar PTB prosthesis with Pelite liner: transtibial amputees' preferences. *JPO: Journal of Prosthetics and Orthotics*, 8(3), 96–99.

Bruno, T. R., & Kirby, R. L. (2009). Improper use of a transtibial prosthesis silicone liner causing pressure ulceration. *American Journal of Physical Medicine & Rehabilitation*, 88(4), 264–266.

Cluitmans, J., Geboers, M., Deckers, J., & Rings, F. (1994). Experiences with respect to the ICEROSS system for transtibial prostheses. *Prosthetics and Orthotics International*, 18(2), 78–83.

Coleman, K. L., Boone, D. A., Laing, L. S., Mathews, D. E., & Smith, D. G. (2004). Quantification of prosthetic outcomes: elastomeric gel liner with locking pin suspension versus polyethylene foam liner with neoprene sleeve suspension. *Journal of Rehabilitation Research & Development*, 41(4), 591–602.

Dillingham, T. R., Pezzin, L. E., MacKenzie, E. J., & Burgess, A. R. (2001). Use and satisfaction with prosthetic devices among persons with trauma-related amputations: a long-term outcome study. *American Journal of Physical Medicine & Rehabilitation*, 80(8), 563–571.

Eshraghi, A., Abu Osman, N. A., Karimi, M. T., Gholizadeh, H., Ali, S., & Wan Abas, W. A. B. (2012a). Quantitative and qualitative comparison of a new prosthetic suspension system with two existing suspension systems for lower limb amputees. *American Journal of Physical Medicine & Rehabilitation*, 91(12), 1,028–1,038.

Eshraghi, A., Abu Osman, N. A., Gholizadeh, H., Karimi, M., & Ali, S. (2012b). Pistoning assessment in lower limb prosthetic sockets. *Prosthetics and Orthotics International*, 36(1), 15–24.

Eshraghi, A., Osman, N. A. A., Gholizadeh, H., Ahmadian, J., Rahmati, B., & Abas, W. A. B. W. (2013a). Development and evaluation of new coupling system for lower limb prostheses with acoustic alarm system. *Scientific Reports*, 3, 1–5.

Eshraghi, A., Abu Osman, N. A., Gholizadeh, H., Ali, S., Sævarsson, S. K., & Wan Abas, W. A. B. (2013b). An experimental study of the interface pressure profile during level walking of a new suspension system for lower limb amputees. *Clinical Biomechanics*, 28(1), 55–60.

Eshraghi, A., Abu Osman, N. A., Gholizadeh, H., Ali, S., & Wan Abas, W. A. B. (2015). Interface stress in socket/residual limb with transtibial prosthetic suspension systems during locomotion on slopes and stairs. *American Journal of Physical Medicine & Rehabilitation*, 94(1), 1–10.

Eshraghi, A., Abu Osman, N.A., Karimi, M., Gholizadeh, H., Soodmand, E., Abas, W. A. B. W. (2014) Gait biomechanics of individuals with transtibial amputation: effect of suspension system. *PLoS ONE*, 9(5), e96988.

Gholizadeh, H., Abu Osman, N. A., Eshraghi, A., Ali, S., & Yahyavi, E. S. (2013). Satisfaction and problems experienced with transfemoral suspension systems: a comparison between common suction socket and Seal-In liner. *Archives of Physical Medicine & Rehabilitation*, 94(8), 1,584–1,589.

Grevsten, S. (1978). Ideas on the suspension of the below-knee prosthesis. *Prosthetics and Orthotics International*, 2(1), 3–7.

Hatfield, A., & Morrison, J. (2001). Polyurethane gel liner usage in the Oxford Prosthetic Service. *Prosthetics and Orthotics International*, 25(1), 41–46.

Kapp, S. (1999). Suspension systems for prostheses. *Clinical Orthopaedics and Related Research*, 361, 55–62.

Kark, L., & Simmons, A. (2011). Patient satisfaction following lower-limb amputation: the role of gait deviation. *Prosthetics and Orthotics International*, 35(2), 225–233.

Kristinsson, Ö. (1993). The ICEROSS concept: a discussion of a philosophy. *Prosthetics and Orthotics International*, 17(1), 49–55.

Kutz, M., Adrezin, R. S., Barr, R. E., Batich, C., Bellamkonda, R. V., Brammer, A. J., Dolan, A. M. (2003). *Standard handbook of biomedical engineering and design*. New York, NY: McGraw-Hill Professional.

McCurdie, I., Hanspal, R., & Nieveen, R. (1997). Iceross—a consensus view: a questionnaire survey of the use of Iceross in the United Kingdom. *Prosthetics and Orthotics International*, 21(2), 124–128.

Narita, H., Yokogushi, K., Shi, S., Kakizawa, M., & Nosaka, T. (1997). Suspension effect and dynamic evaluation of the total surface bearing (TSB) transtibial prosthesis: a comparison with the patellar tendon bearing (PTB) transtibial prosthesis. *Prosthetics and Orthotics International*, 21(3), 175–178.

Schmalz, T., Blumentritt, S., & Jarasch, R. (2002). Energy expenditure and biomechanical characteristics of lower limb amputee gait: the influence of prosthetic alignment and different prosthetic components. *Gait & Posture*, 16(3), 255–263.

SOLIDWORKS [Computer software]. (2009). Retrieved from https://www.solidworks.com/

Smith, D. G., Michael, J. W., & Bowker, J. H. (2004). *Atlas of amputations and limb deficiencies: surgical, prosthetic and rehabilitation principles*. Rosemont, IL: American Academy of Orthopaedic Surgeons.

Van der Linde, H., Hofstad, C. J., Geurts, A. C., Posterna, K., Geertzen, J. H., & Van Limbeek, J. (2004). A systematic literature review of the effect of different prosthetic components on human functioning with a lower-limb prosthesis. *Journal of Rehabilitation Research & Development*, 41(4), 557–570.

Van de Weg, F., & Van der Windt, D. (2005). A questionnaire survey of the effect of different interface types on patient satisfaction and perceived problems among transtibial amputees. *Prosthetics and Orthotics International*, 29(3), 231–239.

Webster, J. B., Chou, T., Kenly, M., English, M., Roberts, T. L., & Bloebaum, R. D. (2009). Perceptions and acceptance of osseointegration among individuals with lower limb amputations: a prospective survey study. *JPO: Journal of Prosthetics and Orthotics*, 21(4), 215–222.

Wirta, R. W., Golbranson, F. L., Mason, R., & Calvo, K. (1990). Analysis of below-knee suspension systems: effect on gait. *Journal of Rehabilitation Research & Development*, 27(4), 385–396.

# 3 Prosthetic Suspension System

*N A Abu Osman*

University of Malaya, Kuala Lumpur, Malaysia

*H Gholizadeh*

Iran University of Medical Sciences (IUMS), Tehran, Iran

## CONTENTS

## 3.1 INTRODUCTION

Transtibial amputations are more common than other amputation types. In recent decades, this degree of amputation has gained a great deal of attention in training and education for surgery, recovery and prosthetics. However, some people with transtibial amputation do not recover entirely (Smith et al., 2004). Surgeons consider three conditions when performing an effective transtibial (below-knee) amputation: maximal bone duration, adequate soft-tissue padding and precise location of nerve endings. The tissue pattern or muscle padding to cover the distal flap is usually the subject of surgical techniques for limb amputation. The end result should be a residual limb with a cylindrical form, strong distal tibial padding, and stable muscles (Smith et al., 2004).

According to research results, suspension and prosthetic fit are closely linked to functional performance and comfort levels (Beil, et al., 2002; Eshraghi et al., 2012). Poor suspension can jeopardize walking pattern, residual limb soft tissue and skin, and comfort (Eshraghi et al. 2012; Gholizadeh et al., 2012a, b; Smith et al., 2004). Suspension devices and sockets are the most important elements of the prosthesis since they make direct contact with the amputee's residual limb. The suspension system should avoid excessive translation, rotation, and vertical motions between the residual limb and socket (Baars and Geertzen, 2005; Klute et al. 2011). Objective and

DOI: 10.1201/9781003196730-3

subjective tests have shown that using silicone liners improves suspension compared to other devices on the market, most likely due to a strong union between the socket and the residual limb (Kristinsson, 1993). However, according to the literature, there is no single suspension device that meets the needs of all people with amputation.

The silicone liner can be attached to prosthetic devices in a variety of ways, such as the lock system (Eshraghi et al. 2013). When designing prosthetic suspension, the following factors should be taken into account: protection, comfort, function, ease of donning/doffing, durability, cosmetic appearance and cost. To address some of the shortcomings of current prosthetic suspension systems, the authors created a new device called HOLO (using hook and loop fabric as a lock system) to be used with silicone liners, which are commercially available and widely used (Gholizadeh et al., 2014a, b). The male portion is sometimes referred to as the anchor, while the female portion is referred to as the loop.

The aim of this study was to implement and test a newly developed prosthetic suspension device in terms of shear strength and patient comfort. Furthermore, shear power, amputee satisfaction and cost comparisons were made between the new system, suction (Seal-In x5 liner), locking (dermo liner), and magnetic MPSS systems.

## 3.2 DESIGNING A NEW SUSPENSION SYSTEM (HOLO)

When designing a prosthetic suspension, the most important factors to consider are protection, comfort, function, ease of donning/doffing, durability, cosmetic appearance and cost. With these considerations in mind, the new system was created using silicone liners, which are commercially available and widely used.

Hook and loop (Velcro) were used as the main part of this suspension system (as a lock system). Two small openings (medial and lateral) were made in the socket wall, in the proximal and distal sections of the socket (Figure 3.1). To prevent any limitations in knee flexion, the proximal opening in the transtibial socket was created below the knee centre. Both openings must be parallel. The hook fastener (polyester hook and loop Velcro V-STRONG, 100%) was used on the socket wall (rolling belt), whereas the loop fastener was attached to the soft liner (silicone liner) (Figure 3.1). A tiny hook (3cm$^2$) was also connected to the distal end of the socket.

The new suspension system was tested mechanically (Figure 3.2) before it was tested on the subjects. Mechanical testing under tensile loading was performed using the universal testing machine INSTRON 4466 to determine how much tensile force each suspension system (lock mechanism) could tolerate before it fails. Furthermore, the other suspension systems used were tested for comparison with the new design.

The study was approved by the Medical Ethics Committee, University of Malaya Medical Centre. Nine transtibial amputees participated in the study. Following the acquisition of written informed consent, each participant was provided with four transtibial prostheses (pin/lock, Seal-In, magnetic (MPSS), and the HOLO suspension system). To ensure a consistent prosthetic quality, fabrication (Figure 3.3) and aligning were done by a single prosthetist. All the subjects were fitted with a transparent check socket to insure that the TSB. They were asked to walk with their new prostheses in the prosthetic laboratory (Department of Biomedical Engineering, University of Malaya, Malaysia) to become familiar with and adapt to the new sockets (Figure 3.4). All the subjects were given a trial period of at least four weeks (for each suspension systems) to become accustomed to the new prostheses.

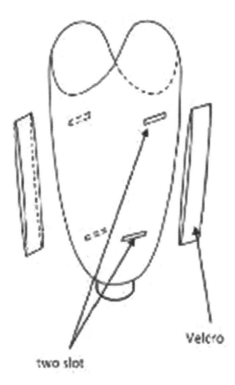

**FIGURE 3.1** Position of Velcro on the socket walls.

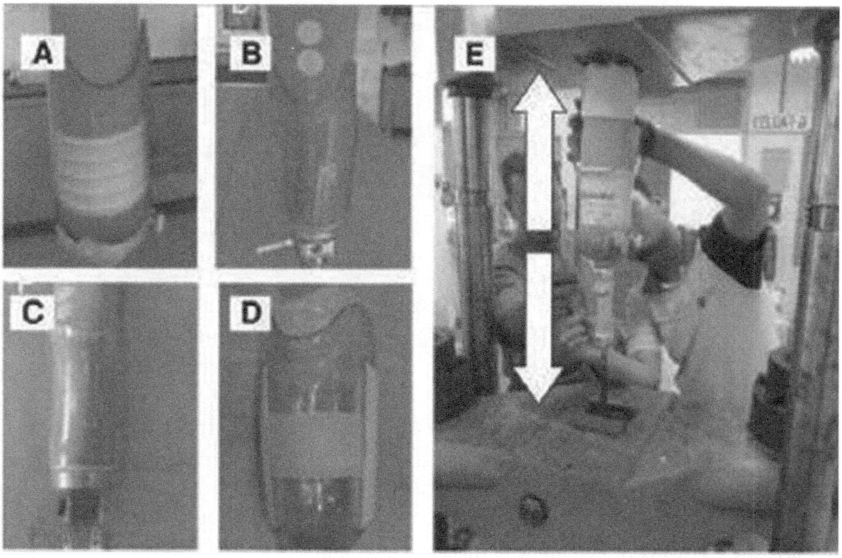

**FIGURE 3.2** Seal-In liner (A); Pin/Lock system (B); Magnet (C); Hook/Loop (D); Tensile testing machine (E).

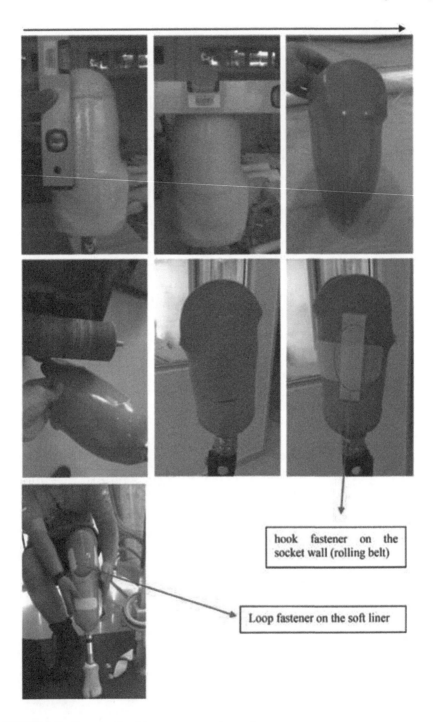

hook fastener on the socket wall (rolling belt)

Loop fastener on the soft liner

FIGURE 3.3   Process of making new suspension system.

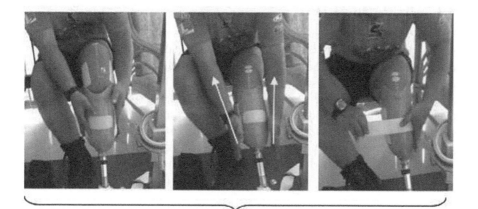

Donning

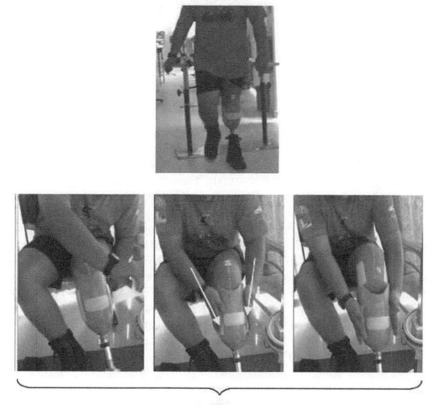

Doffing

**FIGURE 3.4** Donning and doffing process of the new system.

## 3.3   MECHANICAL TEST

The new system could bear a maximum tensile load of 490N (SD, 5.5). Movement within the socket was only 4mm (between the liner and the socket) during the 30 seconds of tensile loading. The pin/lock system could tolerate loading of 580N (SD, 8.5); however, the lock system lost its function after three trials. The MPSS and Seal-In (suction) could tolerate loads of 350.9 (SD, 7) and 310N (SD, 8.4), respectively. With the pin/lock and magnetic system, there was no movement between the end of the liner and socket, and there were 18 and 12mm of traction in the silicone liner, respectively. Furthermore, 7mm of movement between the liner and socket with the Seal-In liner was observed before the system failed.

## 3.4   SUBJECT CHARACTERISTIC

The subjects in this study were all males. Diabetes and trauma were the common causes of amputation, and the mean age (year) and height (cm) of the participants were 42.2 (SD, 14.7) and 174.1 (SD, 7.2) respectively (Table 3.1, Table 3.2). On average, the participants went through amputation 9.7 (SD, 7.5) years prior to the study. The average mass of prostheses (transtibial) for the magnetic (MPSS) suspension, pin/lock (Icelock 200 Series Clutch 4H 214), suction (Seal-In x5), and the new HOLO system among the nine transtibial subjects were 1.89, 1.80, 1.65, and 1.60 kg, respectively.

Participants were generally pleased with the new system, and there was no significant difference with the pin/lock ($P < 0.643$) and magnetic systems ($P < 0.672$). However, there was a significant difference between the new system and suction system using the Seal-In liner ($P < 0.000$). There was no significant difference between the HOLO and the other systems regarding the sitting, walking (even and uneven surface), climbing the stairs, sweating, swelling and smell. The suction suspension system (Seal-In liner) could create better fit compared to the other systems, and there was significant difference between the HOLO and suction systems ($P < 0.002$). Our subjects were happier with the new system (HOLO) due to easy donning and doffing procedures ($P < 0.002$). Also, there was significant difference with the pin/lock and suction systems ($P < 0.000$). The questionnaire revealed that the amputees experienced less pistoning or vertical movement, rotation, and sound inside the socket with the suction socket (Seal-In) compared to the other systems. However, the new system (HOLO) creates more sound compared to the other systems. The irritating sound with the new system was only created during doffing the prosthesis (tearing noise from the hook and loop). The respondents also mentioned that the prosthesis with the suction (Seal-In) made them feel the artificial limb as a natural body part (subjective feedback). Moreover, there was no traction or pain at the distal liner.

**TABLE 3.1**

**Characteristics of the Participants**

| Subject No. | Age | Height (cm) | Mass (kg) | Level of Amputation | Cause of Amputation | Tine since Amputation | Stump Length(cm) | Mobility Grade | Stump Appearance and Problem with lwn Prosthesis |
|---|---|---|---|---|---|---|---|---|---|
| 1 | 39 | 170 | 65 | TT | Trauma | 5 | 14 | K4 | Bony and conical in shape. The bony end of the residual limb was painful during the swing phase of gait. He was using pin/lock system prior to the study. |
| 2 | 23 | 167 | 82 | TT | Trauma | 3 | 15 | K3 | Cylindrical in shape. He was using PTB socket with Pelite (soft liner). He encountered numerous problems with prosthesis, such as pain, wound at the end of his stump, and too much movement (pistoning) within the socket. Most of the weight was centralized at the end of the socket. |
| 3 | 51 | 172 | 67 | TT | Trauma | 5 | 14 | K3 | Bony and conical in shape. The bony end of the residual limb and fibular head were painful during the swing phase of gait and while wearing the prosthesis. He was using pin/lock system prior to the study. |
| 4 | 40 | 180 | 95 | TT | Diabetes | 7 | 16 | K2 | Cylindrical in shape. He was using pin/lock system prior to the study. He encountered difficulties in aligning the pin while wearing the prosthesis. He experienced a disorder in his left hand. |
| 5 | 75 | 182 | 75 | TT | Diabetes | 8 | 13 | K2 | Bony and conical in shape. The bony end of the residual limb was painful during the swing phase of gait. He was using pin/lock system prior to the study. |

*(Continued)*

**TABLE 3.1  (Continued)**
**Characteristics of the Participants**

| Subject No. | Age | Height (cm) | Mass (kg) | Level of Amputation | Cause of Amputation | Tine since Amputation | Stump Length(cm) | Mobility Grade | Stump Appearance and Problem with lwn Prosthesis |
|---|---|---|---|---|---|---|---|---|---|
| 6 | 45 | 185 | 84 | TT | Trauma | 26 | 12 | K3 | Short stump. He was using PTB socket with Pelite (soft liner). He had pain at the end of slump and too much movement (pistoning) within the socket. Most of his weight was centralized at the end of the socket. |
| 7 | 41 | 173 | 95 | TT | Trauma | 5 | 14 | K3 | Cylindrical in shape. He was using pin/lock system prior to the study. He did not have any problem with his prosthesis. |
| 8 | 34 | 175 | 78 | TT | Trauma | 10 | 28 | K3 | Cylindrical in shape. He did not feel any pain at the slump. He was using pin/lock system prior to the study. |
| 9 | 32 | 163 | 72 | TT | Trauma | 18 | 25 | K2 | Conical in shape. Bony prominence was evident at the end of his stump. He did not feel any pain at the stump. He was using pin/lock system prior to the study. |

**TABLE 3.2**

**A Compilation of the Subjective Feedbacks of the Participants**

| Subject no. | Subject's Preference | | Mobility Grade | Subjective Feedback |
|---|---|---|---|---|
| 1 | Seal-In | 1 | K4 | He did not feel any pain at the distal of his residual limb with the Seal-In and the new suspension system during walking. He gained more confidence and also stated that the Seal-In was more suitable than the other suspension systems. Despite that it is more challenging to remove the prosthesis, he still preferred to use the Seal-In system. |
| | Pin/Lock | 4 | | |
| | Magnetic | 2 | | |
| | Hook/Loop | 3 | | |
| 2 | Seal-In | 4 | K3 | He was more satisfied with the silicone liners compared to the PTB with Pelite liner. After changing to silicone liner (TSB socket), he did not have any pain at the distal end of the residual limb, and the wound was healed after two weeks. He felt more confident with the silicone liner and different lock systems (pin/lock, magnet or HOLO). Among the four systems in this study, he preferred the HOLO. the magnetic system, and the pin/lock system. |
| | Pin/Lock | 3 | | |
| | Magnetic | 2 | | |
| | Hook/Loop | 1 | | |
| 3 | Seal-In | 4 | K3 | He did not feel any pain at the distal of residual limb with the Seal-In and the new suspension system. However, he had pain during donning and doffing with the Seal-In liner. He stated that the Seal-In was more suitable during walking, but wearing and removing the prosthesis was extremely more difficult compared to the other suspension systems. |
| | Pin/Lock | 3 | | |
| | Magnetic | 1 | | |
| | Hook/Loop | 2 | | |
| 4 | Seal-In | 4 | K2 | It was very difficult to use the Seal-In due to upper limb weakness. He preferred the hook and loop, pin/lock, and magnetic systems mostly because of their easy donning and doffing. |
| | Pin/Lock | 2 | | |
| | Magnetic | 3 | | |
| | Hook/Loop | 1 | | |
| 5 | Seal-In | 4 | K2 | He did not feel pain with the Seal-In and the new suspension system. Nevertheless, he preferred the new suspension system because of its advantages of easy donning and doffing. He was not happy with the tearing noise during doffing of the prosthesis. |
| | Pin/Lock | 3 | | |
| | Magnetic | 2 | | |
| | Hook/Loop | 1 | | |

*(Continued)*

**TABLE 3.2** (*Continued*)
**A Compilation of the Subjective Feedbacks of the Participants**

| Subject no. | Subject's Preference | | Mobility Grade | Subjective Feedback |
|---|---|---|---|---|
| 6 | Seal-In | 4 | K3 | Pain at the end of the socket was less with the TSB socket compared to the PTB socket. He was satisfied with the pin/lock, hook/loop, and magnetic systems, whereas he felt more socket fit and less rotation inside the socket with the Seal-In. He mentioned that he is not going to use the Seal-In because of the difficulty in donning and doffing. |
| | Pin/Lock | 1 | | |
| | Magnetic | 3 | | |
| | Hook/Loop | 2 | | |
| 7 | Seal-In | 3 | K3 | He felt more socket fit and higher confidence with the Seal-In during walking, however, he was not satisfied with the donning and doffing procedures. He preferred to use the pin/lock and magnetic systems. He was not happy with the hook/lop system because of the sound developing during doffing of the prosthesis. |
| | Pin/Lock | 1 | | |
| | Magnetic | 2 | | |
| | Hook/Loop | 4 | | |
| 8 | Seal-In | 4 | K3 | He was happier with the pin/lock and HOLO systems because of the easy donning and doffing procedures. He also felt less traction at the end of the socket with HOLO and Seal-In system. |
| | Pin/Lock | 1 | | |
| | Magnetic | 3 | | |
| | Hook/Loop | 2 | | |
| 9 | Seal-In | 4 | K2 | He felt more comfortable at the distal end with the Seal-In and the new suspension system, and he was more confident during walking. Regarding the donning and doffing, he preferred the pin/lock and HOLO system. He chose the pin/lock as his first choice because of its easy donning and doffing. |
| | Pin/Lock | 1 | | |
| | Magnetic | 3 | | |
| | Hook/Loop | 2 | | |

## 3.5 SUMMARY AND CONCLUSIONS

Amputee rehabilitation is a challenging procedure which requires expertise, specifically in the selection of prosthetic components based on the amputee's needs. Good prosthetic suspension systems must secure the residual limb inside the prosthetic socket and make donning and doffing procedures easier. The new suspension system is a good alternative for individuals with transtibial amputation as it could solve some problems with the current systems. This system may have some advantages for amputees, including the ease of donning/doffing, firm attachment to the socket, low weight and low cost. The use of Velcro as suspension system might facilitate the donning of prosthesis and reduce traction at the end of the residual limb during the swing phase of gait. The new coupling system proved compatible with the pin/lock system in terms of suspending the leg and the amputee's satisfaction.

## 3.6 DIRECTION FOR FUTURE RESEARCH

This present study proposed a promising suspension system for lower limb amputees. Further research is needed to evaluate more amputees (upper and lower limbs) and to prepare a guideline for the selection of the suspension system. Moreover, sweat control was found to be a major concern with the available prosthetic soft liners. The donning and doffing procedure for soft liners is also problematic for some users, particularly those with upper limb weakness.

## 3.7 ACKNOWLEDGEMENTS

This research was supported by Össur (Reykjavik, Iceland) and the Malaysia UM/ MOHE/HIR grant (project no: D000014-16001). The authors wish to extend their best gratitude to Mr Knut Lechler, Professor Þorvaldur Ingvarsson, Dr Kristleifur Kristjánsson Mr Egill Egilsson, and Ms Ása Guðlaug Lúðvíksdóttir for their kind assistance and technical support in this project.

## REFERENCES

Baars, E., & Geertzen, J. H. B. (2005). Literature review of the possible advantages of silicon liner socket use in transtibial prostheses. *Prosthetics and Orthotics International*, 29(1), 27–37. doi:10.1080/17461550500069612

Beil, T. L., Street, G. M., & Covey, S. J. (2002). Interface pressures during ambulation using suction and vacuum-assisted prosthetic sockets. *Journal of Rehabilitation Research & Development*, 39(6), 693–700.

Eshraghi, A., Abu Osman, N. A., Gholizadeh, H., Karimi, M., & Ali, S. (2012). Pistoning assessment in lower limb prosthetic sockets. *Prosthetics and Orthotics International*, 36(1), 15–24.

Eshraghi, A., Abu Osman, N. A., Gholizadeh, H., Ahmadian, J., Rahmati, B., & Wan, A.W.A.B. (2013). Development and evaluation of new coupling system for lower limb prostheses with acoustic alarm system. *Scientific Reports*. doi:10.1038/srep02270

Gholizadeh, H., Abu Osman, N. A., Kamyab, M., Eshraghi, A., Lúvíksdóttir, A., & Wan Abas, W. A. B. (2012a). Clinical evaluation of two prosthetic suspension systems in a bilateral transtibial amputee. *American Journal of Physical Medicine & Rehabilitation*, 91(10), 894–898.

Gholizadeh, H., Abu Osman, N., Kamyab, M., Eshraghi, A., Wan Abas, W. A. B., & Azam, M. (2012b). Transtibial prosthetic socket pistoning: Static evaluation of Seal-In®X5 and Dermo® Liner using motion analysis system. *Clinical Biomechanics*, 27(1), 34–39.

Gholizadeh, H., Abu Osman, N. A., Eshraghi, A., Ali, S., & Razak, N. (2014a). Transtibial prosthesis suspension systems: systematic review of literature. *Clinical Biomechanics*, 29(1), 87–97.

Gholizadeh, H., Abu Osman, N. A., Eshraghi, A., Ali, S., & Razak, N. (2014b). Transfemoral prosthesis suspension systems: a systematic review of the literature. *American Journal of Physical Medicine and Rehabilitation*, 93(9), 809–823.

Klute, G. K., Berge, J. S., Biggs, W., Pongnumkul, S., Popovic, Z., & Curless, B. (2011). Vacuum-assisted socket suspension compared with pin suspension for lower extremity amputees: effect on fit, activity, and limb volume. *Archives of Physical Medicine and Rehabilitation*, 92(10):1,570–1,575. doi:10.1016/j.apmr.2011.05.019

Kristinsson, Ö. (1993). The ICEROSS concept: a discussion of a philosophy. *Prosthetics and Orthotics International*, 17(1):49–55. doi:10.3109/03093649309164354

Smith, D. G., Michael, J. W., & Bowker, J. H. (2004). *Atlas of amputations and limb deficiencies: surgical, prosthetic, and rehabilitation principles*. Rosemont, IL: American Academy of Orthopaedic Surgeons.

# 4 Motion Analysis for Deep Tendon Reflex

## N A Abu Osman and L K Tham
University of Malaya, Kuala Lumpur, Malaysia

## CONTENTS

## 4.1 INTRODUCTION

The nervous tissues which form the nervous system occupy about 2% of the human body. The system innervates most body parts, playing a major role as receptor, effector and in the coordination of body mechanism. The system functions in a wide area such as regulating body condition, reacting with external stimulus and defending any pathologic change. Due to the vast coverage of nervous tissue in human body, neurological abnormalities will always appear as the first sign of many diseases (Campbell, 2005). The clinical diagnosis of neurological abnormalities is done through procedure named "neurologic investigation".

The advance development of medical devices especially the imaging devices such as magnetic resonance imaging (MRI) and positron emission tomography (PET) improved the quality of neurologic investigation by providing visible evidence to support the diagnosis results, but those tools must be applied at the exact location of lesion. Moreover, images obtained using imaging devices may be a sign of neurological disorder despite of disease suspected by physicians because there are chances where other pathologies which will be the factors of the same defect in the body. Even symptoms such as migraine headaches cannot be detected by imaging devices (Schwartzman, 2006).

An effective neurologic investigation should begin with history taking to diagnose, neurologic examination to prove the diagnosis, and lastly, the imaging

DOI: 10.1201/9781003196730-4

technique and laboratory tests to assist in the planning of treatment procedures (Schwartzman, 2006). History taking includes a series of interviews involving questions such as the patient's personal details, complaints on current illness, the past and present health history (Bickley and Szilagyi, 2003). The neurologic history enables physician to make diagnosis before any physical examination takes place. By obtaining information from the past and current health status, physician will be able to make brief conclusion on the cause of symptoms (Campbell, 2005).

Neurologic examination plays an important role in the process of neurologic investigation. The general aspects which are concerned in the neurologic examination include mental status, cranial nerves, motor system, sensory system, reflexes, cerebellar function and gait (Campbell, 2005). The examination is the way to identify the location of lesion which leads to neurological disorder (Schwartzman, 2006). The mechanism which causes the disease could also be known from neurologic examination (Smith, 2007). Combination of results from different tests provides a complete picture on the diagnosis. This step is vital in neurologic diagnosis as an accurate examination leading to exact determination of therapy procedure which will subsequently increase the probability of healing. Errors in neurologic examination might cause the healing process to be retarded (LeBlond et al., 2008).

Among the aspects of neurologic examination reflexes assessment is especially significant in the investigation of motor system (Toft et al., 1989). The deep tendon reflex that is commonly examined in reflex assessment, is an essential element in the diagnosis of some neurological or neuromuscular disorders (Bishop et al., 1968; Rico and Jonkman, 1982; De Weerd and Jonkman, 1986; Toft et al., 1989; Schott and Koenig, 1991; Frijns et al., 1997; Johnsen and Fuglsang-Frederiksen, 2000; Karandreas et al., 2000). Tendon reflexes, particularly the patellar and Achilles reflexes, are very useful in evaluating the functional disturbance of either a normal or augmented reflex arc (Kuruoglu and Oh, 1993; Karandreas et al., 2000). Clinically, an increased reflex response indicates a neurological disorder which affects the upper motor neuron. On the other hand, a diminished reflex response shows that the disorder is affecting the lower motor neuron (Toft et al., 1989).

Stroke is a common example of disorder affecting the upper motor neuron. Studies conducted in Singapore found the prevalence of stroke to be 4.05%. The rate increased to 7.67% for adults over 65 years old (Venketasubramanian et al., 2005).

According to the World Health Organization (WHO), the global prevalence of diabetes was approximately 2.8% in year 2000, but is estimated to be 4.4% for the year 2030. The number of diabetic patients worldwide has been predicted to increase from 171 million in the year 2000 to 366 million in the year 2030 (Wild et al., 2004). Diabetic neuropathy is the most common complication developed from diabetes persisting high rate of morbidity and mortality (Vinik and Mehrabyan, 2004). It is the condition where the nerve function degrades gradually in a diabetic patient (Boulton and Malik, 1998). This disease affects the lower motor neuron which leads to foot ulcers, gangrene, and most seriously, amputation (Vinik and Mehrabyan, 2004). Studies carried out within Malaysia in 1997 showed about 58% of diabetic patients suffered from diabetic neuropathy (Mustaffa et al., 1998).

Due to the serious consequences of such disorders, the role of reflex assessment cannot be overemphasized. This chapter reviews the significance of deep tendon reflexes, some common methods in clinical reflex assessment, and current issues encountered by the methods described. Existing work in reflex quantification are thoroughly discussed. Lastly, the technique of motion analysis is introduced and its feasibility in quantifying deep tendon reflex is investigated in this chapter.

### 4.1.1   DEEP TENDON REFLEX

The deep tendon reflex which also known as the muscle stretch reflex, is elicited when stretch stimuli is applied to tendon or occasionally to bone, joint, fascia or aponeurotic structures (Campbell, 2005). The reflex is an involuntary reaction of muscle spindles due to a sudden stretch on the tendon (Porter and Weiner, 2004). Reflex response is showed as immediate muscle contraction once the tendon of the related muscle is tapped by a reflex hammer (Walker, 1990; Campbell, 2005).

The deep tendon reflexes were first introduced by Erb and Westphal in 1875 (Lanska, 1989). The method was soon found to be useful, and thus became a famous routine among physicians. Examination on deep tendon reflexes is generally used to evaluate the functionality of central nervous system and peripheral nervous system (Marshall and Little, 2002).

Clinical examination on the deep tendon reflex is an important step to evaluate the motor system (Toft et al., 1989). Increased or absent reflex response indicates a lesion along the reflex pathway. Hyperactive reflex shows an interruption in the area above the reflex arc (Walker, 1990). For this situation, lesion occurs in the upper motor neuron which causes the central nervous system disease such as stroke and spinal cord tumour (Marshall and Little, 2002). On the other hand, hyporeflexic reflex is obtained when the lesion is affecting the lower motor neuron (Walker, 1990). It is often the first sign of peripheral nervous system disease, such as peripheral neuropathy, radiculopathy and lumbar spinal stenosis (Walker, 1990; Marshall and Little, 2002).

Deep tendon reflex is significant in providing useful information for the localization of lesion along the reflex pathway. For instance, normal responses are obtained for biceps and brachioradialis reflexes, but triceps reflex is absent, and all lower reflexes are hyperactive indicating a lesion at C6 to C7 level (Walker, 1990). The deep tendon reflex is also helpful in assessing the effect of training and ageing on human (Kamen and Koceja, 1989; Koceja and Kamen, 1988; Koceja et al., 1991). It can also be used to evaluate the effect of therapeutically induced alteration in the reflex pathway (Delwaide and Pennisi, 1997). In terms of amputation, reflex assessment is commonly deployed as a measure of neurological functionality for pre-surgery planning and preparation and post-surgery evaluation (Isakov et al., 1992).

### 4.1.2   CLINICAL ASSESSMENT OF DEEP TENDON REFLEX

Generally, direct examination could not be done on most parts of the nervous system. One hand will normally be placed on the responding muscle in order to feel the muscle contraction. The degree of a certain reflex is judged by the briskness of the response including the range of movement at the respective joint and the time of muscle contraction. By visually observing and feeling the reflex, examiner grades the response base on knowledge and experience (Campbell, 2005).

Clinical judgement on reflex is done by grading the response using a reflex grading scale (Campbell, 2005). The Mayo Clinic Scale and the National Institute of Neurological Disorders and Stroke (NINDS) scale are two standard scales commonly used in grading reflex responses. The Mayo Clinic Scale is a nine-point scale (Bastron et al., 1956). Stam and van Crevel (1990) evaluated the reliability of Mayo Clinic Scale found that significant disagreement occurred within examiners in the assessments of tendon reflexes. The scale made up of nine grades was too detailed which highly increased the frequency of disagreement in grading reflex responses (Stam and van Crevel, 1990). Thus, the scale was unable to provide reliable prediction for reflexes assessment.

NINDS proposed another myotatic reflex scale in 1993 (Hallett, 1993). The scale was developed with the hope to be accepted as the universal scale in reflexes assessment (Litvan et al., 1996). This is because the utilization of a universal grading scale would greatly reduce errors in clinical judgement of reflex responses (Manschot et al., 1998). The NINDS scale used in-tendon reflex assessment. The scale is similar to some other existing reflex scales in order for it to be easily accepted as the universal scale (Hallett, 1994). Studies conducted by Litvan et al. (1996) found that the intraobserver reliability of the NINDS scale was substantial to near perfect indicating there was great agreement in examiners themselves who assessed reflexes by the scale. Meanwhile, the interobserver reliability found in the study, which was the agreement compared between different examiners, was moderate to substantial. Results from the study confirmed the reliability of the NINDS scale and the potential of the scale to be accepted as the universal reflex scale (Litvan et al., 1996).

A common issue with reflex assessment is the absence of a standard notation in grading reflex responses. According to a survey done in the Netherland, about 20 different reflex scales were used by medical practitioners within the country alone in 1989 (Manschot et al., 1998). The situation is even more significant worldwide. Problems arise when physicians from different backgrounds are judging the same case. There will be great variation in the diagnosis when different scales are used to evaluate a response. There will also be a high probability in conveying wrong information among examiners. Frequent error in reflex assessment affects the quality of neurologic examination and patient management (Manschot et al., 1998).

In a more recent study by Manschot et al. (1998), the performance of Mayo Clinic scale and the NINDS myotatic reflex scale were evaluated and compared. Manschot et al. performed the study under actual clinical setup with the usual examination technique. Low interobserver agreement in both scales was obtained in the study. The reliability found for both scales was also unacceptable. Such great disagreements for the grading scales that require rating selection were due to different opinions and experiences of the medical practitioners from different background (Manschot et al., 1998). The interpretation of reflex briskness by visual observation and feeling the contraction with hand is too examiner dependent. The available grading systems for deep tendon reflex were proved to be varied among examiners leading to difficulties in coordination of therapy programmes. Such situation gives rise to other alternatives seeking for possibilities in interpreting reflex response in a more reliable and systematic way.

### 4.1.3 DEEP TENDON REFLEX QUANTIFICATION

The subjective evaluation of deep tendon reflex leads to extensive research with the aim to assess reflex response in a more objective and quantitative way. Earlier research work involved the quantification of the output of tendon reflex which normally represented by reflex responses. Attempts have also been made to standardize the input which is the tendon tap exerted by reflex hammer.

In the 1970s, Simons and Lamonte developed a solenoid-type reflex hammer operated under electronics control system. The automatic hammer delivered constant taps to the patellar tendon according to preference. Reflex response was measured by Electromyography (EMG) attached to the quadriceps muscle. The force produced in reflex was also measured by a tensiometer linked to the ankle. All data was processed by an automated computer system to produce results of trials immediately (Simons and Lamonte, 1971).

Burke et al. evaluated the effect of Jendrassik manoeuvre on patellar tendon reflex in different age groups. Taps were delivered to the tendon automatically by an electromagnetic solenoid hammer. A force transducer was attached on the ankle to measure the force of reflex responses (Burke et al., 1996).

In 1996, Van de Crommert et al. developed an automated hammer controlled by a magnetic motor. The equipment was fixed to the back of lower leg in order to elicit biceps femoris tendon reflexes during gait. Reflex responses were measured as EMG in this study (Van de Crommert et al., 1996). The same concept was later adopted by Faist et al. to study quadriceps tendon jerk reflex during gait (Faist et al., 1999).

In 1997, Huang et al. proposed a reflex quantification method similar to the study of Simons and Lamonte. The proposed method used a linear actuator that exerted tap to the patellar tendon automatically. The torque sensor was used to measure the reflex torque (Huang et al., 1997).

Cozens et al. built a reflex quantification system consisting of a solenoid reflex hammer controlled by a control system. The reflex hammer driven by a motorized carriage applied taps perpendicular to the biceps tendon. Reflex responses were collected and analyzed as EMG. The experimental setup for the study. The developed system was intended to examine reflex response of patients with head injury who are confined to bed. The device helps monitor the condition of patients by identifying any deterioration as the early sign of secondary brain damage (Cozens et al., 2000).

Other than automated reflex quantification system, many research groups developed reflex hammer operated manually to investigate the properties of tendon taps. Stam and van Crevel studied the Achilles tendon reflexes, patellar tendon reflexes, triceps tendon reflexes and biceps tendon reflexes using a normal reflex hammer. The hammer was attached to a piezoelectric transducer which produced a signal equivalent to the deceleration when the hammer tapped on the tendon. EMG signals were also collected to assess reflex responses (Stam and van Crevel, 1989). Frijns et al. conducted a similar study in 1997 by utilizing a reflex hammer made of conductive rubber. The hammer was connected to a trigger unit for the observation of tapping time. Output of patellar tendon reflex was presented as the EMG amplitude (Frijns et al., 1997). Both studies focused on the effect of different parameter such as height and leg length to reflex latencies and amplitudes in normal subjects.

In 2004, Péréon et al. from France developed an instrumented reflex hammer to study deep tendon reflexes in children and adults. The developed hammer resembled the clinical reflex hammer. The mass located within the handle of hammer will hit the tube wall during tendon tap. This completed an electrical circuit and subsequently produced signal to the data acquisition system. Reflex responses were again presented as EMG signals (Péréon et al., 2004).

Some studies of neurological disorders attempted to quantify both input and output which also involved manual instrumented reflex hammer. Lebiedowska and Fisk conducted a study regarding the reflex and muscular activity of spastic children in 2003. The group used a manual reflex hammer with a strain gauge accelerometer to test on the patellar tendon reflex. A force transducer attached on the ankle functioned to measure the reflex responses during the tests (Lebiedowska and Fisk, 2003). Lebiedowska et al. applied the similar technique to study the patellar tendon reflex in 2011. An electrogoniometer was used to measure the amount of jerk together. Ultrasound imaging was also used to measure the velocities of muscle during the reflex (Lebiedowska et al., 2011).

Mamizuka et al. studied the relationship between tapping force and the angular velocity of leg in patellar tendon reflex for normal and spastic subjects. In the study, a force sensor in the tip of reflex hammer measured the tapping force exerted to the tendon. Reflex response was analyzed as the angular velocity by an accelerometer attached at the ankle (Mamizuka et al., 2007).

LeMoyne et al. produced a reflex quantification device in 2007. The swing arm of the device enabled taps of different forces to be applied on the patellar tendon. The hammer raised to different angles possesses different potential energy, indicating different striking forces. A 3-dimensional accelerometer was placed on the lower leg to quantify reflex responses. The preliminary stage of the study involved two normal subjects and a subject with chronic hemiplegic (LeMoyne et al., 2008).

Generally, attempts to quantify tendon taps involved the development of automated tapping devices and manual instrumented reflex hammers. For automated tapping devices, the position of subject is normally fixed to align the hammer so that taps could be delivered to the exact position of tendon. Manual instrumented hammers are similar to clinical reflex hammer with additional sensors to transmit signals proportional to the force, deceleration or speed of tendon taps.

The automated tapping device is developed to produce constant taps at desired force. Such kind of tapping is predicted to minimize the variation of stimulus delivered to the tendon. However, the experimental setups for automated tapping devices were quite bulky and hard to be applied clinically (Zhang et al., 1999). The fixations on subjects might lead to discomfort and affect the natural response of a reflex (LeMoyne et al., 2008). Stam and Tan investigated the relationship between types of reflex hammer and reflex response in 1987. According to the study, there was no significant difference in reflex response evoked by hand-held reflex hammer and automated reflex hammer. The authors concluded that both methods were equally reliable in providing tendon taps (Stam and Tan, 1987).

Surface EMG is most commonly applied as the method to study reflex responses. The application of EMG to measure reflex responses is very convenient but problem frequent arises in the placement of electrode (Stam and van Crevel, 1990). Moreover, results obtained from these studies still exhibit great variations as it is for clinical reflex assessment (Simons and Dimitrijevic, 1972; Stam and van Crevel, 1989). However, a study conducted by Stam and van Crevel in 1990 found that surface EMG is a reliable method to examine reflex (Stam and van Crevel, 1990) since the variation of reflex response is unavoidable (Lim et al., 2009). Other methods were also applied to assess reflex responses in an objective way, such as force transducer, torque sensor, accelerometer and even ultrasound imaging (Lebiedowska and Fisk, 2003; Mamizuka et al., 2007; LeMoyne et al., 2008; Lebiedowska et al., 2011).

## 4.2 APPLICATION OF MOTION ANALYSIS IN REFLEX QUANTIFICATION

Motion analysis is a technique of comparing sequential still images captured from a body in motion (Griffiths, 2006). The technique is applied in a many research to study human locomotion, providing kinematics and kinetics measurements of joints during movements in gait, running or even sports activities. The technique, which functions to track object in motion and convert the object to a 3D structure (Aggarwal and Cai, 1999), has great potential for analyzing motion of body parts; especially human joints during a reflex. The movement of the reflex hammer tapping on the tendon could be analyzed using this technique.

This section presents a study that proposed a new alternative to assessing patellar tendon reflex using the technique of motion analysis. Quantification of patellar tendon reflex was performed with varying tapping angles as the input, whereas the reflex response captured by an optical motion analysis system, was presented as the study output.

## 4.3 TAPPING VELOCITY

The Queen Square reflex hammer (Figure 4.1) was used to examine patellar tendon reflexes in the study. Three reflective markers (14mm diameter) were attached on the reflex hammer; one was positioned 5cm below the tip, one at half the length of the handle and one on the rubber ring of the hammer head. Using a screw, a protractor was fixed to the tip of the reflex hammer in order to measure the tapping angle.

**FIGURE 4.1** Queen Square reflex hammer used in current study.

The trajectories of the markers attached to the reflex hammer were obtained from the experiments to calculate the velocity components $v_x$, $v_y$ and $v_z$. The average velocities of the hammer's head at all captured frames were calculated using the formula $v = \sqrt{v_x^2 + v_y^2 + v_z^2}$. The experimental maximum velocity was compared to the velocities calculated using the theory of conservation of mechanical energy, as $v = \sqrt{2gL(1 - \cos\theta)}$ where the gravitational acceleration $(g) = 9.81 \text{m/s}^2$, the handle length of the reflex hammer (meter) and $\theta$ is the tapping angle in degree (°).

In this study, patellar tendon reflex will be quantified and studied using six tapping angles: 15°, 30°, 45°, 60°, 75° and 90°. Table 4.1 demonstrates the mean values of maximum velocities at every tapping angle exerted in reflex tests.

The maximum tapping velocities of different tapping angles obtained in experiments were compared statistically. According to Table 4.2 the experimental tapping velocities were statistically different for all tapping conditions. Significant difference was found for all the comparison pairs of maximum velocity at different tapping angles.

The tapping angle was found to be linearly related to the experimental maximum velocity, as shown in Figure 4.2. The coefficient of determination, $r^2$ was close to 1 ($r^2 = 0.989$). This shows that the maximum velocity increases significantly as the tapping angle increases.

In general, the reliability of the tapping method was supported by (1) a linear relationship between the tapping angle and the experimental maximum tapping velocity; (2) small variances in experimental maximum velocity elicited in any tapping angle; and (3) non-significant differences between experimental and theoretical maximum velocities. The results also proved the consistency and reproducibility of the method in producing tendon taps at different tapping angles. Constant and reproducible tendon taps is the key factor to improve the reliability and sensitivity of reflex assessment while reducing the variability of results obtained (Marshall and Little, 2002). Moreover, the method is a simple one, which does not require instrumented tapping device that involve complicated setup and high cost (Zhang et al., 1999). Tapping device with force sensor is not required with the proposed method, the use of ordinary reflex hammer exerted taps that will usually be practiced in clinical assessment. Clinicians could examine reflexes as usual without feeling unfamiliar as

---

**TABLE 4.1**

**Mean and One Standard Deviation Values of Experimental Maximum Tapping Velocities at Different Tapping Angles**

| Tapping Angle (°) | Mean (m/s) | Standard Deviation |
|---|---|---|
| 15 | 0.466 | 0.059 |
| 30 | 0.918 | 0.066 |
| 45 | 1.345 | 0.051 |
| 60 | 1.758 | 0.064 |
| 75 | 2.130 | 0.061 |
| 90 | 2.472 | 0.077 |

## TABLE 4.2
## Statistical Comparisons of Experimental Maximum Tapping Velocities between Different Tapping Angles

| Comparison Pair | | P value |
|---|---|---|
| 15° | 30° | < 0.001* |
| | 45° | < 0.001* |
| | 60° | < 0.001* |
| | 75° | < 0.001* |
| | 90° | < 0.001* |
| 30° | 45° | < 0.001* |
| | 60° | < 0.001* |
| | 75° | < 0.001* |
| | 90° | < 0.001* |
| 45° | 60° | < 0.001* |
| | 75° | < 0.001* |
| | 90° | < 0.001* |
| 60° | 75° | < 0.001* |
| | 90° | < 0.001* |
| 75° | 90° | < 0.001* |

* Statistical significance difference.

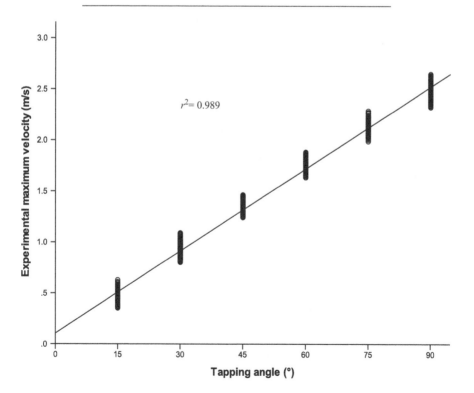

**FIGURE 4.2** Representative relationship between the tapping angle and the experimental maximum tapping velocity.

in custom made instrumented reflex hammer. In addition, there was no feedback of unpleasant from the subjects with the method of tapping. This method could be implemented easily in clinical reflex assessment to monitor the precision of tendon taps applied to the tendons.

## 4.4 PATELLAR TENDON REFLEX QUANTIFICATION

The effect of varying tapping intensity was studied by assessing the patellar tendon reflex at both left and right sides with tapping angles of 15°, 30°, 45°, 60°, 75° and 90°. The patellar tendon reflex response was interpreted as the range of motion of knee extension during a reflex. The knee angles in reflex tests were collected in $x$-, $y$- and $z$-direction by the technique of motion analysis. However, only knee angles of the $x$-axis were used for analysis in the study due to the nature of patellar tendon reflex movement. The knee angle at $x$-axis indicates movement of the sagittal plane which includes flexion-extension of the joint. On the other hand, $y$-axis represents motion in the frontal plane including the abduction–adduction of the joint whereas $z$-axis represents motion in the transverse plane which is the internal–external rotation of knee. The major movements of lower leg during patellar tendon reflex are flexion and extension. Movements of knee joint in the frontal and transverse planes are not significant. Therefore, knee angles in the sagittal plane were the main concern and documented as the reflex amplitude.

Figure 4.3 shows the mean reflex amplitudes obtained for 100 subjects where tendon taps were applied to both left and right legs at different tapping angles. The reflex amplitude was found to increase gradually with the increasing tapping angle.

On the other hand, data involving all subjects were assigned to two random groups, with 50 subjects in each group, in order to examine the consistency of reflex output. The statistical comparisons of reflex amplitude between the random groups as shown in Figure 4.4 found no significant difference.

The relationship between reflex input which was measured as maximum tapping velocity and reflex amplitude was further analyzed and displayed in Figure 4.5. The Pearson's correlation coefficient of knee angle and maximum tapping velocity was strong ($r$ value = 0.501) with statistically significant $P$ value ($<0.001$).

Patellar tendon reflex amplitude is shown to be quantified successfully according to tapping angle in the current study. This simple method shows a great potential in shifting clinical reflex assessment from examiner-dependent to a more objective manner. The significant differences of reflex amplitude between different tapping angles showed that the current method is sensitive in detecting output variation according to tapping input. Significant difference in reflex amplitude was noted between tapping angle of 30° and 45°, and 45° and 60°, suggesting that 45° or 60° is the best tapping angle to elicit an adequate response in clinical practice. Although larger tapping angles produced greater output, the increment of output with tapping angles larger than 60° was minimal and not significant. According to the study by Zhang et al. in 1999, excessive tapping force might cause the tendon to counteract the respective tissues and thus would not develop great reflex response as expected.

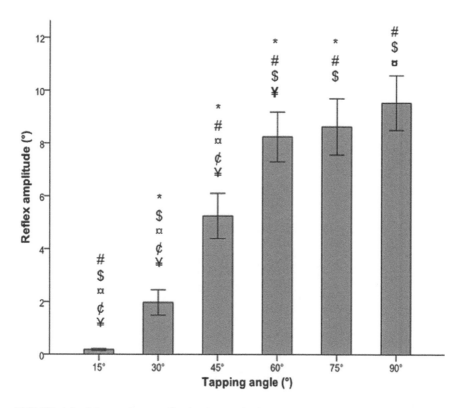

**FIGURE 4.3** Mean reflex amplitudes (± standard error) grouped under different tapping angles. The reflex amplitude increased gradually with the increment of tapping angle. * denotes statistical significance difference compared to tapping angle of 15°. # denotes statistical significance difference compared to tapping angle of 30°. $ denotes statistical significance difference compared to tapping angle of 45°. ¤ denotes statistical significance difference compared to tapping angle of 60°. ¢ denotes statistical significance difference compared to tapping angle of 75°. ¥ denotes statistical significance difference compared to tapping angle of 90°. Symbols (¤, ¥) in bold indicates statistical significance difference at $P < 0.01$, symbols not in bold indicates statistical significance difference at $P < 0.001$.

The reflex amplitude was found to increase gradually with the increase of tapping angle. A small tapping angle does not produce sufficient stimulus to the tendon and there would be little reflex response (Zhang et al., 1999). Knee angles collected at a tapping angle of 15° was mainly due to the movements of leg by the subject. A larger tapping angle generates greater stimulus on the muscle receptors that would ease the development of action potentials (Zhang et al., 1999). This explains the increasing trend of reflex amplitude with the increase of tapping angle. The linear relationship between tapping input and reflex amplitude proved the validity of this method in measuring reflex output. The finding also agrees with some earlier studies where the reflex responses are linearly related to the reflex inputs (Marshall and Little, 2002; Zhang et al., 1999).

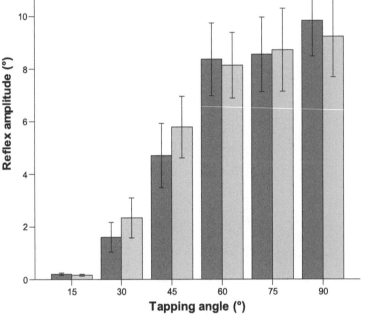

**FIGURE 4.4** Mean reflex amplitudes (± standard error) of random assigned groups according to tapping angle.

In neurological examination, the intra- and inter-observer variability of tendon reflex responses is well known to be large (Stam and Tan, 1987; Stam and van Crevel, 1990). Even with the development of accurate and efficient reflex quantification tools or methods in clinical studies, reflex output is still showing large variability than predicted (Simons and Dimitrijevic, 1972; Stam and Tan, 1987; Lim et al., 2009; Tham et al., 2011). The same observation was obtained in the present study, where the large standard deviations indicate a large range of reflex amplitudes at all tapping inputs. The variation in reflex responses can be affected by psychological factors such as the mental condition and anxiety of the subject, and physiological factors such as the position and muscular activity of subject (Stam and Tan, 1987; Stam and van Crevel, 1990).

The tapping angle of 45°, as the common tapping range applied to the tendon in clinical reflex assessment (Tham et al., 2011), elicited relatively small reflex responses in the study. According to the results, the mean reflex amplitude of 5.25° obtained by tapping with 45° could hardly be noticed through visual observation. This increases the difficulty of judgement on reflex briskness by the clinicians. Reflex quantification methods will enable the clinicians to grade the reflexes more accurately.

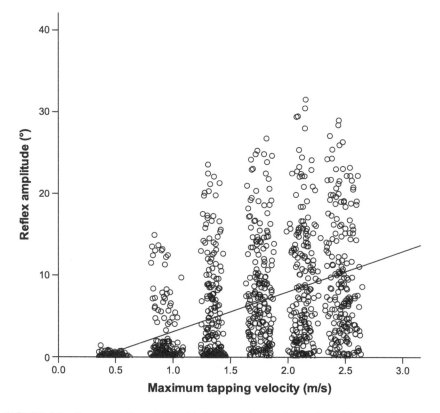

**FIGURE 4.5**   Representative relationship between the tapping angle and knee angle.

## 4.5   CONCLUSION

This chapter presents the importance of deep tendon reflex assessment in facilitating clinical diagnosis and treatment implementation. The subjective evaluation of deep tendon reflex leads to extensive research with the aim to assess reflex response in a more objective and quantitative way. Motion analysis has been studied and the results proved its reliability to objectively quantify patellar tendon reflex, suggesting the feasibility of applying the technique as a bedside tool in clinical quantification of deep tendon reflexes. Application of new technologies in assisting objective clinical assessment increases the quality of care while providing valuable information to help better understanding of neurological disorders.

## REFERENCES

Aggarwal, J. K., Cai, Q., (1999). Human motion analysis: a review. *Computer Vision and Image Understanding* 73, 428–440.

Bastron, J. A., Bickford, R. G., Brown, J. R., et al. *(Mayo Clinic Staff)*. (1956). *Clinical examinations in neurology*. Philadelphia, PA: Saunders.

Bickley, L. S., Szilagyi, P. G. (2003). *Bates' guide to physical examination and history taking* (8th ed.). Philadelphia, PA: Lippincott Williams & Wilkins.

Bishop, B., Machover, S., Johnston, S., (1968). Role of gamma motor system in the Achilles tendon reflex of hemiplegic patients. *Archives of Physical Medicine and Rehabilitation* 49, 698–707.

Boulton, A. J., Malik, R. A. (1998). Diabetic neuropathy. *Medical Clinics of North America* 82, 909–929.

Burke, J. R., Schutten, M. C., Koceja, D. M., Kamen, G., (1996). Age-dependent effects of muscle vibration and the Jendrassik maneuver on the patellar tendon reflex response. *Archives of Physical Medicine and Rehabilitation* 77, 600–604.

Campbell, W. W. (2005). *DeJong's the neurologic examination* (6th ed.). Philadelphia, PA: Lippincott Williams & Wilkins.

Cozens, J. A., Simon, M., Chambers, I. R., Mendelow, A. D., (2000). Monitoring of head injury by myotatic reflex evaluation. *Journal of Neurology, Neurosurgery and Psychiatry* 68, 581–588.

De Weerd, A. W., Jonkman, E. J., (1986). Measurement of knee tendon reflex latencies in lumbar radicular syndromes. *European Neurology* 25, 304–308.

Delwaide, J., Pennisi, G., (1997). A comparative electrophysiologic study of diazepam and tetrazepam in patients with spasticity. *Journal of Neurological Rehabilitation* 11, 91–96.

Faist, M., Ertel, M., Berger, W., Dietz, V., (1999). Impaired modulation of quadriceps tendon jerk reflex during spastic gait: differences between spinal and cerebral lesions. *Brain* 122, 567–579.

Frijns, C. J., Laman, D. M., van Duijn, M. A., van Duijn, H., (1997). Normal values of patellar and ankle tendon reflex latencies. *Clinical Neurology and Neurosurgery* 99, 31–36.

Griffiths, I. W. (2006). *Principles of biomechanics and motion analysis*. New York: Lippincott Williams & Wilkins.

Hallett, M., (1993). NINDS myotatic reflex scale. *Neurology* 43, 2,723.

Hallett, M., (1994). Myotatic reflex scale. *Neurology* 44, 1984.

Huang, H., Zhang, L., Rymer, W. Z. (1997). *A computer-controlled electromechanical hammer to quantify tendon reflex. Paper presented at the 19th Annual International Conference of the IEEE Engineering in Medicine and Biology Society*, Chicago, IL.

Isakov, E., Susak, Z., Korzets, A. (1992). Reflex sympathetic dystrophy of the stump in below-knee amputees. *The Clinical Journal of Pain* 8(3), 270–285. doi:10.1097/00002508-199209000-00014. PMID: 1421743.

Johnsen, B., Fuglsang-Frederiksen, A. (2000). Electrodiagnosis of polyneuropathy. *Neurophysiologie Clinique* 30, 339–351.

Kamen, G., Koceja, D. M., (1989). Contraleral influences on patellar tendon reflexes in young and old adults. *Neurobiology Aging* 10, 311–315.

Karandreas, N., Piperos, O., Dimitriou, D., Kokotis, P., Zambelis, T., (2000). Electrophysiological recording of the tendon reflexes in cervical myelopathy. *Electromyography and Clinical Neurophysiology* 40, 83–88.

Koceja, D. M., Burke, J. R., Kamen, G., (1991). Organisation of segmental reflexes in trained dancers. *International Journal of Sports Science* 12, 285–289.

Koceja, D. M., Kamen, G., (1988). Conditioned patellar tendon reflexes in sprint- and endurance-trained athletes. *Medical Science of Sports Exercises* 20, 172–177.

Kuruoglu, R., Oh, S. J., (1993). Quantitation of tendon reflex in normal volunteers. *Electromyography and Clinical Neurophysiology* 33, 347–351.

Lanska, D. J., (1989). The history of reflex hammers. *Neurology* 39, 1,542–1,549.

Lebiedowska, M. K., Fisk, J. R., (2003). Quantitative evaluation of reflex and voluntary activity in children with spasticity. *Archives of Physical Medicine and Rehabilitation* 84, 828–837.

Lebiedowska, M. K., Sikdar, S., Eranki, A., Garmirian, L., (2011). Knee joint angular velocities and accelerations during the patellar tendon jerk. *Journal of Neuroscience Methods*
198, 255–259.

LeBlond, R. F., DeGowin, R. L., Brown, D. D. (2008). *DeGowin's diagnostic examination*
(9th ed.). New York: McGraw-Hill Professional.

LeMoyne, R., Dabiri, F., Jafari, R., (2008). Quantified deep tendon reflex device, second generation. *Journal of Mechanics in Medicine and Biology* 8, 75–85.

Lim, K. S., Bong, Y. Z., Chaw, Y. L., Ho, K. T., Lu, K. K., Lim, C. H., Toh, M. L., Tan, C. T.,
(2009). Wide range of normality in deep tendon reflexes in the normal population.
*Neurology Asia* 14, 21–25.

Litvan, I., Mangone, C. A., Werden, W., Bueri, J. A., Estol, C. J., Garcea, D. O., Rey, R. C.,
Sica, R. E. P., Hallett, M., Bartko, J. J., (1996). Reliability of the NINDS myotatic reflex
scale. *Neurology* 47, 969–972.

Mamizuka, N., Sakane, M., Kaneoka, K., Hori, N., Ochiai, N., (2007). Kinematic quantitation
of the patellar tendon reflex using a tri-axial accelerometer. *Journal of Biomechanics* 40,
2,107–2,111.

Manschot, S., van Passel, L., Buskens, E., Algra, A., van Gijn, J., (1998). Mayo and NINDS
scales for assessment of tendon reflexes: between observer agreement and implications
for communication. *Journal of Neurology, Neurosurgery and Psychiatry* 64, 253–255.

Marshall, G. L., Little, J. W., (2002). Deep tendon reflexes: a study of quantitative methods.
*Journal of Spinal Cord Medicine* 25, 94–99.

Mustaffa, B. E., Wan Mohamad, W. B., Chan, S. P., Rokiah, P., Mafauzy, M., Kumari, S.,
Chandran, A. A., Ong, G. K. C., Jogensen, L. N., Yeo, J. P., for The Diabetes Data
Collection Project (Diabcare – Malaysia) Study Group (1998). The current status of
diabetes management in Malaysia. *Journal of the Asean Federation of Endocrine
Societies* 16(2) Supplement, 1–13.

Péréon, Y., Nguyen The Tich, S., Fournier, E., Genet, R., Guihéneuc, P., (2004).
Electrophysiological recording of deep tendon reflexes: Normative data in children and
in adults. *Clinical Neurophysiology* 34, 131–139.

Porter, N. C., Weiner, W. J. (2004). The neurologic examination. In W. J. Weiner & C. G. Goetz
(Eds.), *Neurology for the non-neurologist* (5th ed., pp. 1–20). Philadelphia, PA:
Lippincott Williams & Wilkins.

Rico, E., Jonkman, E. J., (1982). Measurement of the Achilles tendon reflex for the diagnosis
of lumbosacral root compression syndromes. *Journal of Neurology, Neurosurgery and
Psychiatry* 45, 791–795.

Schott, K., Koenig, E., (1991). T-wave response in cervical root lesions. *Acta Neurologica
Scandinavica* 84, 273–276.

Schwartzman, R. J. (2006). *Neurologic examination* (1st ed.). MA: Blackwell Publishing.

Simons, D. G., Dimitrijevic, M. R., (1972). Quantitative variations in the force of quadriceps
responses to serial patellar taps in normal man. *American Journal of Physical Medicine*
51, 240–263.

Simons, D. G., Lamonte, R. J., (1971). Automated system for the measurement of reflex
responses to patellar tendon tap in man. *American Journal of Physical Medicine* 50,
72–79.

Smith, D. S. (2007). *Field guide to bedside diagnosis* (2nd ed.). Philadelphia, PA: Lippincott
Williams & Wilkins.

Stam, J., Tan, K. M., (1987). Tendon reflex variability and method of stimulation.
*Electromyography and Clinical Neurophysiology* 67, 463–467.

Stam, J., van Crevel, H., (1989). Measurement of tendon reflexes by surface electromyography
in normal subjects. *Journal of Neurology* 236, 231–237.

Stam, J., van Crevel, H., (1990). Reliability of the clinical and electromyographic examination
of tendon reflexes. *Journal of Neurology* 237, 427–431.

Tham, L. K., Abu Osman, N. A., Lim, K. S., Pingguan-Murphy, B., Wan Abas, W. A. B., Mohd Zain, N., (2011). Investigation to predict patellar tendon reflex using motion analysis technique. *Medical Engineering & Physics* 33, 407–410.

Toft, E., Sinkjaer, T., Espersen, G. T., (1989). Quantitation of the stretch reflex. Technical aspects and clinical applications. *Acta Neurologica Scandinavica* 79, 384–390.

Van de Crommert, H. W. A. A., Faist, M., Berger, W., Duysens, J., (1996). Biceps femoris tendon jerk reflexes are enhanced at the end of the swing phase in humans. *Brain Research* 734, 341–344.

Venketasubramanian, N., Tan, L. C. S., Sahadevan, S., Chin, J. J., Krishnamoorthy, E. S., Hong, C. Y., Saw, S. M., (2005). Prevalence of stroke among Chinese, Malay, and Indian Singaporeans: A community-based tri-racial cross-sectional survey. *Stroke* 36, 551–556.

Vinik, A. I., Mehrabyan, A., (2004). Diabetic neuropathies. *Medical Clinics of North America* 88, 947–999.

Walker, H. K. (1990). Deep tendon reflexes. In H. K. Walker, W. D. Hall & J. W. Hurst (Eds.), *Clinical methods: The history, physical and laboratory examinations* (3rd ed.). London: Butterworth Publishers.

Wild, S., Roglic, G., Green, A., Sicree, R., King, H., (2004). Global prevalence of diabetes: estimates for the year 2000 and projections for 2030. *Diabetes Care* 27, 1,047–1,053.

Zhang, L.-Q., Huang, H., Silwa, J. A., Rymer, W. Z., (1999). System identification of tendon reflex dynamics. *IEEE Transactions on Rehabilitation Engineering* 7, 193–203.

# 5 Air Pneumatic Suspension System (APSS)

*N A Abu Osman*

University of Malaya, Kuala Lumpur, Malaysia

*G Pirouzi*

A-9 Sarvestan, Parsa Tower, Koye Sahand, Tabriz, Iran

## CONTENTS

DOI: 10.1201/9781003196730-5

## 5.1  INTRODUCTION

### 5.1.1  AMPUTATION

Major causes of amputations include non-neutralized bombs and land mines from past wars; traffic accidents, particularly motorcycle accidents (Ebskov, 1992); and diabetes (Adler et al., 1999). Over the years, amputation incidents have not decreased because of accidents, wars, and chronic diseases (Dillingham et al., 2002, Bader D). Amputations, particularly of the lower limb, are increasing in developed countries for various reasons (NHS Scotland, 2004–2005). The first recorded successful amputation dates to 484 BC, when Hegesistratus reportedly escaped from prison by cutting off one of his feet. The earliest limb amputations generally resulted in death because of blood loss or infection. Amputations were previously referred to as "guillotine" operations because all tissues were divided in one stroke, often by blade, at the same level.

Other medical advances that help reduce the hazards of amputations involve the use of ligatures or sutures (originally described by Hypocrites and reintroduced by Paré in 1529) to cauterize blood vessels and the use of tourniquets (Silver-Thorn, 2003).

The first transtibial prosthesis, a lower limb prosthesis, includes a thigh corset and external hinges introduced in 1696 by Verduyn and by Potts in 1816. Amputation surgeries have been significantly improved using anaesthetics, such as ether and chloroform, and antiseptics. James Edward Hanger opened the first artificial limb store in the United States during the Civil War in 1861.

Other historical transtibial prosthetic developments include the patellar tendon bearing (PTB) prosthesis (Radcliffe and Foort, 1961), total surface bearing (TSB) (Kristinsson, 1993) and hydrostatic interface design (HST) (Figure 5.1).

There is a considerable demand for lower limb prostheses globally because of vascular diseases, diabetes, motorbike accidents and natural disasters. The wars in the Middle East, which are affecting thousands of people, have given rise to the number of amputees.

According to the data collected from 125,094 Malaysian employees who have undergone social secretary organization (SOCSO), the prevalence of diabetes has

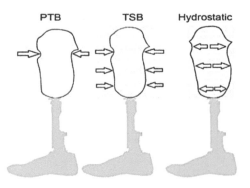

**FIGURE 5.1**   Three basic categories of transtibial sockets (arrows indicate main force vectors).

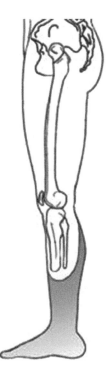

**FIGURE 5.2**   Below-knee amputation.

reached alarming levels. The World Health Organization projects that by 2030, approximately 2.48 million people in Malaysia will have diabetes.

In the United States, nearly 2 million people live with limb loss, and 113,000 lower limb amputations are performed each year (Brain, 2013). The number of people living with limb loss is projected to double by the year 2050 (Ziegler-Graham et al., 2008).

Below-knee (BK) amputation is known as transtibial amputation, which is the amputation of the shin bone (Figure 5.2). BK amputation is the most common type of amputation, and the risk of postoperative complications is considerably lower than that of transfemoral amputation. In BK amputation, the knee joint is safe, and walking with prosthesis is normally more successful (Silver-Thorn, 2003).

## 5.1.2  PROSTHESIS

The main objective of rehabilitation after lower limb amputation is to restore function gait pattern in amputees. Some evidence suggests that the first prosthesis was created in 300 BC (Table 5.1). It was made of thin pieces of bronze and wood, and the residual limb suspended by a leather skirt (Hanger Inc.).

The science of prosthetics continues to make considerable progress, particularly after World War II when disabled soldiers returned with lost limbs and required medical attention and rehabilitation.

## TABLE 5.1
## Historical Timeline of Prosthetics (Reproduced from (Ron Seymour, 2002))

| | |
|---|---|
| 43,000 BC | Evidence was found that amputation was done with primitive tools. |
| 2730–2625 BC | A device to stabilize the knee joint originating from this period was found. |
| 1500 BC | Indian literature describes artificial legs. |
| 370 BC | Hippocrates used splints on the legs. |
| 485–425 BC | Herodotus described an individual imprisoned by Sparta who supplied himself with a wooden loot. |
| 300 BC | A prosthesis unearthed in the ruins of Pompeii is thought to be the first prosthesis. |
| 131–201 | Galen used dynamic orthoses for scoliosis and kyphosis. |
| 476–1453 | During the Middle Ages, knights wore elaborate armour to conceal prostheses. |
| 1200 | Medical school at Bologna considers orthotics as an important part of medical knowledge. |
| 1509–1590 | Ambroise Paré established technical standards for surgical amputations and described spinal corsets and shoe modifications. |
| 1690 | Verduin constructed a transtibial prosthesis with copper socket, leather thigh corset and a wooden foot. |
| 1790–1847 | Lisfranc, a famous surgeon, amputates a foot in less than one minute. |
| 1800 | Baron Larrey, surgeon to Napoleon Bonaparte, performs 200 amputations on the battlefield in one day. He advocates wounds being operated on within the first 24 hours. |
| 1860 | Mortality rate because of sepsis in London for transtibial and transfemoral amputations were 50% and 80%, respectively. |
| 1865 | Lord Lisler starts surgical antisepsis to decrease high mortality rates. |
| 1865 | J E Hangar sustains an amputation while serving in the Confederate Army, places rubber bumpers in solid feet, and produces the first articulated prosthetic foot. |
| 1918 | The Limb Fitting Centre at Queen Mary's Hospital, Roehampton becomes a primary development and supply centre to military veterans after World War I. |
| 1945 | The US Veterans Administration supports the development of the patellar tendon bearing and the quadrilateral sockets. Canada develops a prosthetic research programme at Sunnybrook Hospital in Toronto. |
| 1970 | The US Veterans Administration develops the endoskeletal prosthesis. |
| 2000 | A microprocessor-controlled knee with hydraulic swing and stance phase control is developed. |

### 5.1.3 TRANSTIBIAL AMPUTATION AND PROSTHESIS

Transtibial amputation is the most common amputation (Table 5.2). The risk of post-operative complications in transtibial amputation is considerably lower than that of transfemoral amputation. Many developments have been achieved on prosthetic components for upper and lower limb prostheses, which employed new microprocessor technology to control prostheses and simulate the performance of the human organs; however, these legs are too expensive for majority of patients.

These systems include the belt and suprapatellar cuff (Radcliff et al., 1961), figure-of-8 belt (Girling and Cummings, 1972), sleeve suspension (Chino et al., 1975), supracondylar–suprapatellar suspension (Breakey, 1973), supracondylar suspension, thigh corset silicon liner suspension, and distal locking pin, lanyard and suction suspension.

**TABLE 5.2**
**Demographics**

| Subjects | Gender | Age | Amputation Causes | Level of Activities | Height (cm) | Mass (kg) | Stump Length (cm) |
|---|---|---|---|---|---|---|---|
| 1 | Male | 36 | Motorbike accident | K3 | 172 | 85 | 15 |
| 2 | Male | 72 | Vascular disease | K2 | 180 | 73 | 13.5 |
| 3 | Male | 50 | Motorbike accident | K3 | 176 | 65 | 13 |
| 4 | Male | 23 | Motorbike accident | K3 | 168 | 60 | 13 |
| 5 | Male | 38 | Vascular disease | K3 | 178 | 99 | 16 |

Limited community ambulatory (K2), Community ambulatory (K3).

The suspension systems in lower limb prostheses have a vital role in prosthetic function (Kristinsson, 1993; Klute et al., 2010; Baars and Geertzen, 2005; Isozaki et al., 2006; Tanner and Berke, 2001). Suspension and fitting play important roles in comfort and prosthetic function (Baars and Geertzen, 2005; Isozaki et al., 2006; Kristinsson, 1993; Tanner and Berke, 2001). Silicon liners were introduced in 1986 and it was claimed they had the main advantage of enhanced bond with the stump; therefore, better suspension than the other soft sockets (Baars et al., 2008). Several systems are employed to secure the stump inside a socket and connect the suspension system to the pylon (adaptor) and the foot. However, silicon and polyethylene foam liners are the most commonly used suspension systems. Vacuum liner (sealed-in) is a solution for this problem. In addition, the most important factor mentioned by the amputees is the fit of their prosthesis and suspension (Datta et al., 1996; Fillauer et al., 1989; Legro et al., 1999).

However available suspension systems come with several problems linked to the continuous change in residual limb size, volume, donning and doffing, especially in systems with vacuum liners (Gholizadeh et al., 2012; Eshraghi et al., 2012). The continuous change in the volume or size of the residual limb leads to loss of contact and interruption of pressure distribution in the entire system (Goswami et al., 2003). Meanwhile, a misaligned pin in the pin-lock suspension system introduces difficulties, such as failure in locking the suspension system to the pylon and blisters on the residual limb of the amputee, when the pin suspension system becomes old and is improperly fitted (Klute et al., 2010). The shuttle-lock system/pin liner also creates a large suction distally on the residual limb and causes chronic skin change (Beil and Street, 2004). Socket, liner and residual limb should be in full contact to ensure proper pressure distribution (Rommers et al., 2000). Pressure is concentrated on some parts of the system, resulting in injury to the residual limb (Gailey et al., 2008).

## 5.2 PROBLEM STATEMENT

Socket shape determines how weight is distributed. Issues related to the control of pressure distribution caused by the continuous change in residual limb size remain unsolved despite attempts to address these limitations. Trying to compensate by angling the pin into the dock can cause soreness on the residual limb of the amputee or cause the pin to be jammed in the dock, making it impossible to remove the prosthesis. The weight of the patient rests on just a few points that contact the socket when socket is not fitted properly.

## 5.3 OBJECTIVE OF THIS STUDY

The aim of my study is to design a suspension system by aid of pneumatic device to overcome residual limb volume changes and to eliminate the problems related to the pin-lock and vacuum locking systems. The main objectives of this study also include donning and doffing and pressure distribution of the transtibial prosthesis.

## 5.4 LITERATURE REVIEW

### 5.4.1 INTRODUCTION

Several publications have shown the qualitative and quantitative analysis using modern measuring technologies based on biomechanical statistical calculations of pressure measurement on the residual limb in contact with socket during weight bearing and non-weight weight bearing phases of gait. Biomechanical knowledge of the mutual behaviour of the stump, the socket, and its attachment leads to the improvement of the functions of the prostheses (Jia et al., 2004). Although prosthetists use modern technologies that consider the biomechanical parameters of people with disabilities (PWDs), these patients remain dissatisfied with prostheses (Dillingham et al., 2002, Portnoy et al., 2008).

The PTB design demonstrated is the concept of selecting concentration weight bearing (Figure 5.3). The PTB socket emerged and became popular around 1957 after several years of using the artificial joint–corset below-knee limb (Radcliffe and Foort, 1961).

The objective of this section is to summarize the literature that addresses the recent studies on prosthetic biomechanics, socket–stump interface, socket fitting, pressure distribution and shear force. To overcome the problems, in 1993 Kristinsson developed the TSB prosthetic socket called Iceross, in which weight is distributed on the entire socket surface (Kristinsson, 1993). A method recently developed by Noroozi et al. enables a prosthetist to analyze the pressure distribution of the prosthetic socket by inverse problem analysis approach (Sewell et al., 2012). The question of whether quantitative evaluations apply in clinical or non-clinical lower limb prostheses remains despite its importance and impact on the performance and satisfaction of the patient with a prosthetic device.

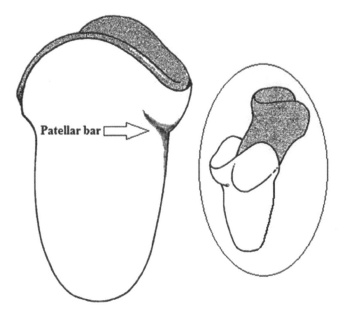

**FIGURE 5.3** Patellar tendon bearing (PTB) socket and soft liner.

### 5.4.2 MULTIDISCIPLINARY APPROACH

Compared with that of other sciences, tables and databases relevant to prosthetics and orthotics are very few, such as the National Amputee Statistical Database, RECAL Information Services, National Prosthetics Patient Database (NPPD, 2013), REHABDATA, CIRRIE (Center for International Rehabilitation Research Information and Exchange), and Able Data. This pressure distribution directly and indirectly affects the effective indices of user satisfaction, issues relevant to stump interaction, and the mechanical and biomechanical variables of the socket. Since 1945 (Vasconcelos, 1945), many studies have measured pressure distribution inside the socket for comparative and biomechanical evaluation and for a better understanding of the stump–socket interface. The skin concurrently has chemical, physical and mechanical factors, which undergo problematic changes when interacting with the socket and the suspension system, causing skin damage. Thus, researchers of prosthetics prefer an interdisciplinary approach because of the direct and indirect relationship of this area with other sciences (Nielsen, 1990). Evaluation and measurement of these changes and the heat around the stump, and the correct understanding of this interaction, are important issues (Peery et al., 2005).

### 5.4.3 UNDERSTANDING STUMP–SOCKET BEHAVIOUR

Findings such as skin damage caused by "a loading cycle (22–118 times) of 4–23 kilopascals with a friction coefficient of 0.5" (Sanders et al., 1992) can support more reliable qualitative results because notions such as endurance threshold, peak point

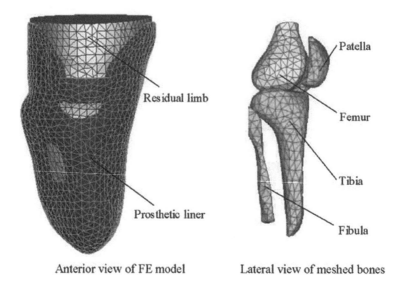

Anterior view of FE model            Lateral view of meshed bones

**FIGURE 5.4**    FEM for prosthetic socket and residual limb (adapted from Ref. Jia et al., 2004).

and the onset of pain are subjective, varying from one person to another. Parameters such as force distribution, friction, and tension on the stump against the socket have been used in finite element method (FEM) (Figure 5.4), which is a suitable instrument for biomechanically assessing sockets. The foundations of material and fluid physics, such as hydrostatics and Pascal's law, can provide a suitable framework for understanding stump and socket behaviour (Laing et al., 2011). Similar to an artificial leg, which is a system that should be considered a whole, a stump is a complicated system in which mechanical and biomechanical behaviour should be considered in a consistent framework. Remarkable ethnic differences also exist in terms of genetics, race, muscle intensity and skin endurance, lowering the credibility of the results of such experiments.

For instance, liner stiffness and its impact on stump–socket friction have been evaluated, with the solid model constructed using auto-mesh function of computer-aided design (CAD) (Pro/Engineering 2000, Parametric Technology Corporation, USA) system (Lin et al., 2004). As in other common methods, several parameters in this modelling are taken as fixed or integrated to prevent the complexity of and interference in results (Figure 5.5).

However, displacement during the donning of the socket is not identified sometimes of modelling because of the incapability of the model.

Software such as Solidworks (Solidworks Corporation, Massachusetts, USA), Abaqus (Hibbitt, Karlsson, and Sorensen, Inc., Pawtucket, RI, USA), MRI, and XRY Dynamic have been widely used to understand the biomechanics of suspension systems and stump interfaces (Figure 5.6). After the model and the data are transferred from Solidworks to Abaqus for FEM, the bones, fat, and muscles are assumed to be the same elements that form a monolith with different mechanical properties (Portnoy et al., 2009).

**FIGURE 5.5** FEM of stump. The mesh at the centre illustrates the bone; two layers in the centre are the soft tissues; and the outer layer represents the liner.

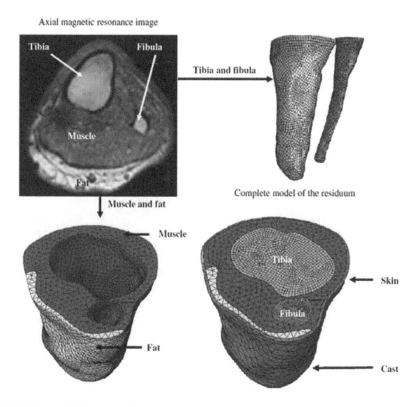

**FIGURE 5.6** FEMs created from axial MRI and the model is analyzed the Abaqus (adapted from Ref. (Portnoy et al., 2009)).

Thus, pain perception was evaluated using technical and computerized analyses to redesign and rectify the socket according to obtained results before its actual construction (Lee and Zhang, 2007). Subsequent studies have highlighted the mutual effect of hard and soft tissues of the stump (e.g., effect of knee movement, changes in the position of stump bones, and their role in generating tension and shear forces). Understanding mutual socket–stump interface using measuring instruments and modelling pressure and tension has been the focus of earlier experiments.

The results were expected to contribute practical improvements to socket construction, indicating that the final aim was to reduce the pain caused by misfit sockets.

The study of stump–socket interface behaviour in a dynamic state by using an assessment instrument has automatically extended the research scope. However, systematic thinking has surfaced in these studies, although some assumptions govern these evaluations, such as considering the bone and stump soft tissue the same.

### 5.4.4 Deep-tissue Damage

Some studies have been conducted on deep-tissue damage (DTD) to understand the major causes of such (Stekelenburg et al., 2008) exerted pressure and weight-induced stress, and the reaction force of the socket–stump, damage and undermine the skin. Portnoy et al. conducted a pioneering study using 2D and 3D modelling and evaluation to prepare a preventive model for DTD in transtibial amputation (Portnoy et al., 2008) (Figure 5.7).

Extensive research using FEM has covered the features of materials used in the socket and liner, the geometric dimensions of the socket–stump, and forces and their effects on the configuration of the artificial leg, shear stress intensity, pressure distribution and friction. The possibility of evaluating and recording data and their exchange in mechanical and biomechanical evaluation systems, such as FEM, is the main index

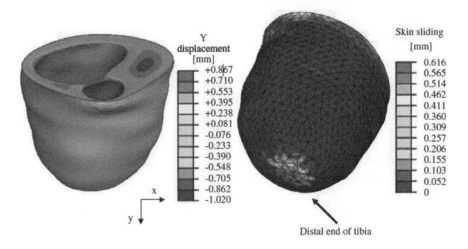

**FIGURE 5.7** FEMs in conditions about 1mm displacement forced upon fibula and tibia bones (adapted from Ref. Portnoy et al., 2008).

of this model of socket design Roith. The skin has a mutual interaction not only with the interface of the socket but also with the tissues, tibia and fibula and muscles.

All femoral movements and their resultant displacements make the pressure point in the stump very difficult to identify.

### 5.4.5  SENSING PRESSURE

Defects of measuring instruments and the weakness of existing sensors in terms of size, sensitivity and the impact of heat, which cast doubt on the research results, have attracted the attention of researchers. Lee and Zhang (2007) aimed to rectify socket design in a virtual environment before its construction by using the pain perception threshold as the evaluation criterion (Figure 5.8).

Measurement systems include the shear stress neuromuscular system, 3D computer modelling, prototype socket sensor matrices, customized pressure vessels, the Rincoe Socket Fitting System, the Tekscan F-Socket pressure measurement system, and the Novel Pliance System. Special sensors are utilized for evaluating the pressure between the socket and the stump and between the liner and the socket. The thickness of these sensors and systems, albeit small, affects the results of studies (Buis and Convery, 1997; Polliack et al., 2000). Modern biofeedback and pressure measurement machines have been used to record effective pressure in dynamic and static states.

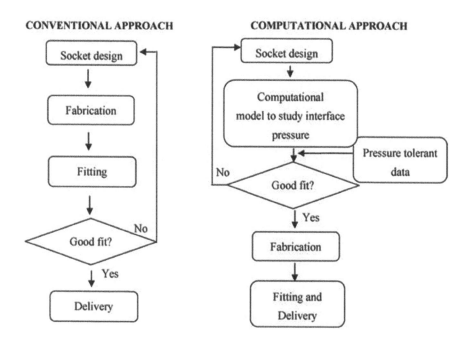

**FIGURE 5.8**  Computer-assisted conventional methods for socket assessment and design (adapted from Ref. Lee and Zhang, 2007).

### 5.4.6 Socket Design and Pressure Distribution

Duplicating a socket (Steele, 1994) seems irrational because a stump changes over time, and the socket can be rectified only by prosthetists (Lilja et al., 1998). Several techniques based on other sciences, including CAD and computer-aided manufacture (CAM), have been used to reduce the fabrication time of sockets. Their results showed the effect of time, the changes in pressure distribution in the stump–socket interface, and the increased risk of injuries in the stump. Such a reduction strategy based on the expensive CAD/CAM should be considered when the demand for artificial legs rapidly increases but improving the production rate does not improve the quality of prosthesis fit (Sewell et al., 2012). Studies have been conducted to overcome these negative consequences by direct skeletal attachment (Figure 5.9) of limb prostheses since the 1980s (Branemark et al., 2001).

Although the final aim of the studies is the satisfaction of PWDs, one of the most important advantages of modelling and instrumental evaluation is that they are independent of user feedback and uncontrollable variables such as patient perception, allowing the creation of systematic data and tables and making the evaluation compliant with the scientific method. However, an evaluation model allows the probe into quantitative and qualitative areas and obtains a more adequate evaluation by using statistical and mathematical methods in the sense that qualitative variables can be tested by if construction variables are fixed and exerting defined controls based on some standardization. Aesthetic parameters should be considered, and its results are regarded as an evaluation index of patient satisfaction. These limitations include the mechanical and biomechanical properties of the bone and its difference with materials, bone behaviour over time, issues such as bone atrophy and osteoporosis, and generated damage, and concerns related to inflammation in the skin surrounding the implant (Pitkin, 2009). Studies have been limited to an increased and improved biomechanical understanding of the interactions among the components of an artificial leg. All existing studies have been conducted in available sockets, and existing logic is likely to change with the changes in performance.

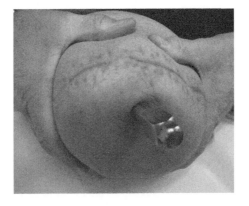

**FIGURE 5.9**  Direct skeletal attachment of limb prostheses (adapted from Ref. Tillander et al., 2010).

## 5.5 METHODOLOGY

This chapter discusses the development of a new APSS by aid of semiconductor pressure sensor to overcome residual limb volume changes, provide better pressure distribution between stump–socket interface, and eliminate problems related to the pin-lock and vacuum locking systems.

### 5.5.1 PARTICULARS OF THE DESIGN

APSS comprises a control board with microcontroller that includes a semiconductor pressure sensor (ADP41410/Panasonic, New Jersey, United States), an air cuff attached inside the socket, air pumps, and pressure-regulating valves (Figure 5.10).

The desired pressure between the socket and the stump is defined by the user and controlled smartly by an APSS Microco.

System operation is controlled and guaranteed through standard protocols of multiple feedbacks by the microprocessor, and pressure data transmission through the module that contains the sensor. To use the new pneumatic system, the amputee first defines the desired pressure before placing the stump into the socket. Finally, the amputee needs to press a button to release the air pressure to doff the prosthesis (Figure 5.11).

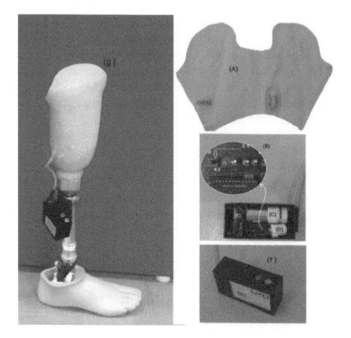

**FIGURE 5.10** Feature and components of the APSS: Bladder (**a**); Control circuit board (including a pressure sensor and microcontroller) (**b**); Pump (**c**); Valve (**d**); Battery (**e**); Operation system (**f**); Assembled transtibial prostheses (**g**).

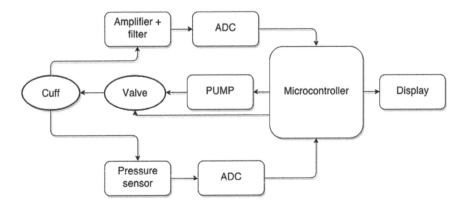

**FIGURE 5.11**   System operation chart.

### 5.5.2   EXPERIMENTAL PROTOCOL

The subjects practiced walking in the motion analysis laboratory under the control of the prosthetist who aligned the prostheses according to their needs. The subjects wore the new system for five hours before attending the motion laboratory for pressure measurements. The patients were asked to stand bearing full weight on the prosthetic limb before the weights were loaded. Pistoning within the socket was monitored to measure the vertical displacement point, beginning with the pressure of 25kPa. Then, pressure and loading weights were applied under non-weight bearing condition. The indicator attached to the liners was a measuring reference for the displacement from the edge of the posterior trim line of the socket. Transtibial prostheses with the Iceross silicon liners were used to evaluate the performance of the APSS, but the pins were removed from the liners (Figure 5.12).

Four pressure sensors were attached, to the anterior, posterior, medial and lateral surfaces of the residual limb. The applied pressure for calibration was the ratio of each subject's body weight to the respective sensor area (Buis and Convery, 1997). The sensor arrays were placed inside the Tekscan PB100T pressure bladder and subjected to a repeated pressure of 100kPa for equilibration under the manufacturer's instructions. The residual limbs were covered with cellophane plastic wrap, and each transducer was attached to the cellophane plastic wrap with spray adhesive (Scotch Super Adhesive, 3M Corporate, St). An F-scan (Tekscan, South Boston, Massachusetts, USA) pressure measurement system was used to evaluate the pressure at all the surfaces (Tekscan, 2010). Each pressure sensor was individually trimmed to fit to the contours of the residual limb. Four consecutive trials were completed by the subjects on the walkway, and approximately eight to nine steps were taken in each trial. Pressure sensors arrays were equilibrated and calibrated to reduce the chances of inaccuracies (Figure 5.13).

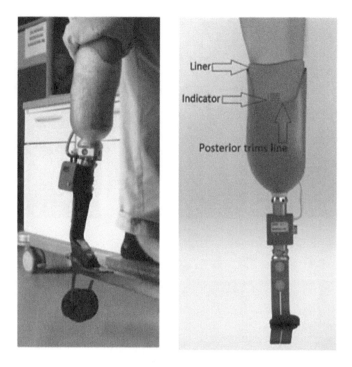

**FIGURE 5.12**   Pistoning within the socket was monitored to measure the vertical displacement point.

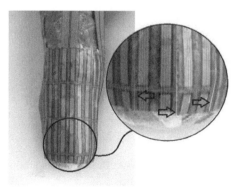

**FIGURE 5.13**   Transducers were divided in parts to fit the contours of liners.

### 5.5.3   PRESSURE DISTRIBUTION BETWEEN SOCKET–STUMP INTERFACE

Pressure distribution between the stump and the new suspension was tested using F-scan sensors, as shown in Figure 5.14 and Figure 5.15 Pressure values were measured in 12 regions of the residual limb.

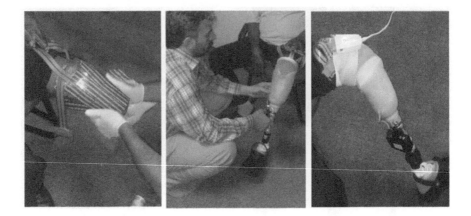

**FIGURE 5.14**   F-scan test setup.

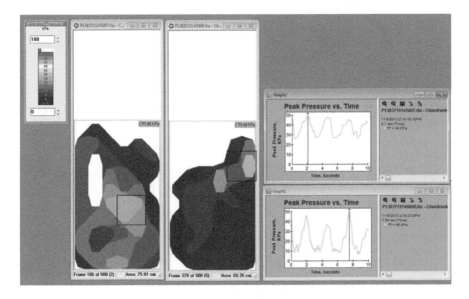

**FIGURE 5.15**   Data and graphs show pressure distribution and the peak points of pressure.

## 5.6   RESULTS

### 5.6.1   Applicability as a Suspension System

The pressure needed to maintain contact between the stump and the artificial limb with no applied external load was approximately 10kPa. External weight was applied on the artificial limb, and pistoning was measured in the different cases. Results are presented in Figure 5.16.

The pressure was read at the distal, middle and proximal portions of the stump in the anterior, posterior, lateral and medial sides. The acceleration factors of the

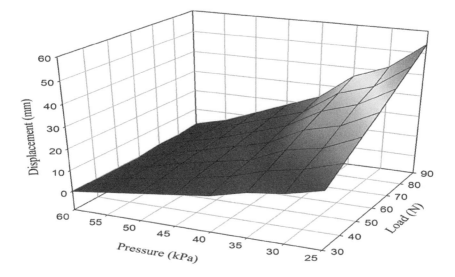

**FIGURE 5.16**  Average values of distances between posterior trims line and liner (mm).

dynamic acceleration tolerance of the system were evaluated by replacing the force loading under static conditions presents the pattern of data points. Average pressure was 54kPa on the anterior and posterior sides and 45.6kPa on the lateral and medial sides (Table 5.3). The results indicates a strong negative relationship between pressure exerted by the inflation bladder and the load applied. Pistoning clearly decreased when pressure was increased by the inflation bladder in the socket–stump interface.

**TABLE 5.3**
**The Average Peak Pressure (kPa) for All Sensors' Sites at the Medial and Lateral Anterior, Posterior, Residual Limb**

| Anterior | | | Posterior | | |
|---|---|---|---|---|---|
| **Proximal** | **Middle** | **Distal** | **Proximal** | **Middle** | **Distal** |
| **(Var)** | **(Var)** | **(Var)** | **(Var)** | **(Var)** | **(Var)** |
| 54.81 | 56.43 | 51.3 | 50.94 | 56.52 | 53.73 |
| (13.21) | (7.69) | (9.56) | (12.41) | (6. 82) | (17.26) |
| | **Medial** | | | **Lateral** | |
| **Proximal** | **Middle** | **Distal** | **Proximal** | **Middle** | **Distal** |
| **(Var)** | **(Var)** | **(Var)** | **(Var)** | **(Var)** | **(Var)** |
| 42.84 | 44.91 | 44.55 | 47.7 | 50.49 | 43.38 |
| (8.83) | (8.91) | (6.23) | (12.05) | (7.52) | (13.25) |

**TABLE 5.4**
**Level of Satisfaction**

| Satisfaction Parameters | Excellent | Good | Average | Poor |
|---|---|---|---|---|
| Fit of prosthesis | 1 | 3 | 1 | |
| Ability to don and doff the prosthesis | 3 | 2 | | |
| Ability to sit with the prosthesis | 3 | 1 | 1 | |
| Ability to walk with the prosthesis | 2 | 3 | | |
| Ability to walk on uneven terrain | 2 | 3 | | |
| Ability to walk up and down on stairs | 2 | 3 | | |
| Suspension | 2 | 1 | 1 | 1 |
| Appearance of the prosthesis | 2 | 3 | | |
| Overall satisfaction with the prosthesis | 1 | 3 | 1 | |
| Overall score | 18 | 23 | 4 | 1 |

### 5.6.2 SATISFACTION

Patients were asked to answer qualitative questions related to the functionality of the new suspension system to determine the level of their satisfaction, and the results are tabulated in Table 5.4.

The overall score indicated that the new system was accepted by the patients and the new approach is applicable. But many factors must be considered, such as the duration of prosthetic use.

## 5.7  DISCUSSION

### 5.7.1 EVALUATION OF THE STATUS QUO

But an accurate and acceptable model has not been developed because of the breadth and plurality of effective parameters in the mutual relationship between the stump and the socket. Studies have been limited to an increased and improved biomechanical understanding of the interactions among the components of an artificial leg. Studying a part of the artificial leg and its interaction with the stump requires the elimination of the effects of other parts. Systematic investigation of an artificial leg requires quantitative and qualitative data and technical tables. Quantitative studies can provide an almost logical and effective framework for consensus building and a common language among experts. Some modelling parameters in the FEM-based assessments are considered fixed or eliminated according to the preference of the model developer. This trend is an evolutionary trend, and the quality of life of PWDs is expected to improve, given technological advances.

### 5.7.2 FOCUSED ON INDICATORS

Studies conducted on the biomechanical behaviour of the stump and the socket interface (Sewell et al., 2000; Zhang and Roberts, 2000; Jia et al., 2004; Goh et al., 2005; Jia et al., 2005; Portnoy et al., 2008; Portnoy et al., 2009) highlight the importance of the fitting. Studies have been recently conducted on the biomechanical behaviour in the stump–socket interface (Sewell et al., 2000; Zhang and Roberts, 2000; Jia et al., 2004; Goh et al., 2005; Jia et al., 2005; Portnoy et al., 2008; Portnoy et al., 2009), and the importance of these research topics is very clear. Another advantage of the APSS system is that prosthesis fitting adjustment is performed after donning and doffing is performed by release pressure. The study introduces a new design of suspension system developed based on innovation and creativity to solve the problems that have not been solved by other kinds of suspension systems yet. Statistical analysis shows this new system has enough advantage to compete with the suspension systems and can replace other systems by adding new powerful and adjustable materials and intelligent electronic systems growing rapidly. System size can be adjusted, and volume changes compensated by the intelligent system by this new suspension. Size can be adjusted in the new suspension system, and volume changes are compensated by the intelligent system. The basis of the concept of APSS is a suspension system and cushioning are its fringe benefits. The mean peak pressure (kPa) at the anterior in medial for APSS is 56.43kPa, whereas in a recent study conducted by Ali et al. Pressure distribution in the socket–stump interface and socket fit, which are the main indicators for evaluating a socket (Kristinsson, 1993; Czerniecki and Gitter, 1996), were considered in this study. In addition, automatic pressure adjustment is a unique capability of the APSS and is not available in systems. The APSS can also provide better fitting because the bladder mimics stump geometry after inflation pressure and lays on it.

## 5.8 CONCLUSIONS

Pressure distribution in socket–stump interface and fitting of a socket are the main indicators in evaluating and assessing a socket. Therefore, these two factors have been evaluated in this study. Imaging techniques were used to evaluate pistoning, and F-scan was used to evaluate pressure distribution. Evaluating the system shows that the achievement of optimum pistoning was solved by the mechanisms used in APSS. However, complete removal of pistoning is neither possible nor desirable. An adjustable socket is important in a variety of daily activities. APSS can maintain suspension performance in different situations with different pressures. APSS makes it possible for a patient to adjust pressure if he/she needs more mobility. The tests performed show that the pressure can be set for normal conditions and can be changed to increase mobility if necessary. Another advantage of the APSS system is the prosthesis fitting adjustment performed after donning and doffing is performed by the release of pressure. Therefore, donning and doffing are easier in the APSS system than in the sockets.

The study introduces a new design of suspension system developed based on innovation and creativity to solve the problems that have not been solved by other kinds of suspension systems. APSS was developed by applying knowledge of intelligent materials and employing smart electronic devices and applications. In future studies, the artificial leg should be considered an integrated ergonomic system and the entire prosthesis should be evaluated in a comprehensive systematic model.

## REFERENCES

Adler, A. I., Boyko, E. J., Ahroni, J. H. & Smith, D. G. (1999). Lower-extremity amputation in diabetes. The independent effects of peripheral vascular disease, sensory neuropathy, and foot ulcers. *Diabetes Care*, 22, 1,029–1,035.

Baars, E. C. T., Dijkstra, P. U. & Geertzen, J. H. B. (2008). Skin problems of the stump and hand function in lower limb amputees: a historic cohort study. *Prosthetics and Orthotics International*, 32, 179–185.

Baars, E. C. T. & Geertzen, J. H. B. (2005). Literature review of the possible advantages of silicon liner socket use in transtibial prostheses. *Prosthetics and Orthotics International*, 29, 27–37.

Beil, T. L., & Street, G. M. (2004). Comparison of interface pressures with pin and suction suspension systems. *Journal of Rehabilitation Research & Development*, 41(6).

Brain, S. (2013). Diabetes amputation prevention [Online]. Statistic Brain Research Institute. Available: http://www.statisticbrain.com/.

Branemark, R., Branemark, P. I., Rydevik, B. & Myers, R. R. (2001). Osseointegration in skeletal reconstruction and rehabilitation: a review. *Journal of Rehabilitation Research and Development*, 38, 175–181.

Breakey, J. (1973). Criteria for use of supracondylar and supracondylar–suprapatellar suspension for below-knee prostheses. *Prosthetics and Orthotics International*, 27, 14–18.

Buis, A. W. & Convery, P. (1997). Calibration problems encountered while monitoring stump/socket interface pressures with force sensing resistors: techniques adopted to minimize inaccuracies. *Prosthetics and Orthotics International*, 21, 179–182.

Chino, N., Pearson, J. R., Cockrell, J. L., Mikishko, H. A. & Koepke, G. H. (1975). Negative pressures during swing phase in below-knee prostheses with rubber sleeve suspension. *Archives of Physical Medicine and Rehabilitation*, 56, 22–26.

Czerniecki, J. M., & Gitter, A. J. (1996). Gait analysis in the amputee: has it helped the amputee or contributed to the development of improved prosthetic components. *Gait & Posture*, 4(3), 258–268.

Datta, D., Vaidya, S. K., Howitt, J. & Gopalan, L. 1996. Outcome of fitting an ICEROSS prosthesis: views of trans-tibial amputees. *Prosthetics and Orthotics International*, 20, 111–115.

Dillingham, T. R., Pezzin, L. E. & Mackenzie, E. J. (2002). Limb amputation and limb deficiency: epidemiology and recent trends in the United States. *Southern Medical Journal*, 95, 875–883.

Ebskov, L. B. (1992). Level of lower limb amputation in relation to etiology: an epidemiological study. *Prosthetics and Orthotics International*, 16, 163–167.

Eshraghi, A., Abu Osman, N. A., Gholizadeh, H., Karimi, M. & Ali, S. (2012). Pistoning assessment in lower limb prosthetic sockets. *Prosthetics and Orthotics International*, 36, 15–24.

Fillauer, C. E., Pritham, C. H., & Fillauer, K. D. (1989). Evolution and development of the silicone suction socket (3S) for below-knee prostheses. *JPO: Journal of Prosthetics and Orthotics*, 1(2), 92–103.

Gailey, R., Allen, K., Castles, J., Kucharik, J. & Roeder, M. (2008). Review of secondary physical conditions associated with lower-limb amputation and long-term prosthesis use. *Journal of Rehabilitation Research and Development*, 45, 15–29.

Gholizadeh, H., Abu Osman, N. A., Kamyab, M., Eshraghi, A., Abas, W. A. B. W. & Azam, M. N. (2012). Transtibial prosthetic socket pistoning: static evaluation of Seal-In (R) X5 and Dermo (R) Liner using motion analysis system. *Clinical Biomechanics*, 27, 34–39.

Girling, J. & Cummings, G. (1972). Artificial-limb fabrication without the use of commercially available components. *Prosthetics and Orthotics International*, 4, 21–25.

Goh, J. C. H., Lee, P. V. S., Toh, S. L., & Ooi, C. K. (2005). Development of an integrated CAD–FEA process for below-knee prosthetic sockets. *Clinical Biomechanics*, 20(6), 623–629.

Goswami, J., Lynn, R., Street, G., & Harlander, M. (2003). Walking in a vacuum-assisted socket shifts the stump fluid balance. *Prosthetics and Orthotics International*, 27(2), 107–113.

Isozaki, K., Hosoda, M., Masuda, T. & Morita, S. (2006). CAD/CAM evaluation of the fit of transtibial sockets for transtibial amputation stumps. *Journal of Medical and Dental Sciences*, 53, 51–56.

Jia, X., Zhang, M. & Lee, W. C. (2004). Load transfer mechanics between transtibial prosthetic socket and residual limb--dynamic effects. *Journal of Biomechanics*, 37, 1,371–1,377.

Jia, X., Zhang, M., Li, X., & Lee, W. C. (2005). A quasi-dynamic nonlinear finite element model to investigate prosthetic interface stresses during walking for trans-tibial amputees. *Clinical Biomechanics*, 20(6), 630–635.

Klute, G. K., Glaister, B. C., & Berge, J. S. (2010). Prosthetic liners for lower limb amputees: a review of the literature. *Prosthetics and Orthotics International*, 34(2), 146–153.

Kristinsson, Ö. (1993). The ICEROSS concept: a discussion of a philosophy. *Prosthetics and orthotics International*, 17(1), 49–55.

Laing, S., Lee, P. V., & Goh, J. C. (2011). Engineering a trans-tibial prosthetic socket for the lower limb amputee. *Annals of the Academy of Medicine-Singapore*, 40(5), 252.

Lee, W. C. C. & Zhang, M. (2007). Using computational simulation to aid in the prediction of socket fit: a preliminary study. *Medical Engineering & Physics*, 29, 923–929.

Legro, M. W., Reiber, G., Del Aguila, M., Ajax, M. J., Boone, D. A., Larsen, J. A., Smith, D. G. & Sangeorzan, B. (1999). Issues of importance reported by persons with lower limb amputations and prostheses. *Journal of Rehabilitation Research and Development*, 36, 155–163.

Lilja, M., Hoffmann, P., & Öberg, T. (1998). Morphological changes during early trans-tibial prosthetic fitting. *Prosthetics and Orthotics International*, 22(2), 115–122.

Lin, C. C., Chang, C. H., Wu, C. L., Chung, K. C., & Liao, I. C. (2004). Effects of liner stiffness for trans-tibial prosthesis: a finite element contact model. *Medical Engineering & Physics,* 26(1), 1–9.

NHS Scotland, 2004–2005. *The amputee statistical database for the United Kingdom.*

Nielsen, C. C. (1990). A survey of amputees: functional level and life satisfaction, information needs, and the prosthetist's role. *Journal of Prosthetics and Orthotics*, 3, 125–129.

NPPD. (2013). *National Prosthetics Patient Database (NPPD)* [Online]. Available: http://www.virec.research.va.gov/NPPD/Overview.htm.

Peery, J. T., Ledoux, W. R. & Klute, G. K. (2005). Residual-limb skin temperature in trans-tibial sockets. *The Journal of Rehabilitation Research and Development*, 42, 147.

Pitkin, M. (2009). On the way to total integration of prosthetic pylon with residuum. *Journal of Rehabilitation Research and Development*, 46(3), 345.

Polliack, A. A., Sieh, R. C., Craig, D. D., Landsberger, S., McNeil, D. R., & Ayyappa, E. (2000). Scientific validation of two commercial pressure sensor systems for prosthetic socket fit. *Prosthetics and Orthotics International,* 24(1), 63–73.

Portnoy, S., Siev-Ner, I., Shabshin, N., Kristal, A., Yizhar, Z. & Gefen, A. (2009). Patient-specific analyses of deep tissue loads post transtibial amputation in residual limbs of multiple prosthetic users. *Journal of Biomechanics*, 42, 2,686–2,693.

Portnoy, S., Yizhar, Z., Shabshin, N., Itzchak, Y., Kristal, A., Dotan-Marom, Y., Siev-Ner, I. & Gefen, A. (2008). Internal mechanical conditions in the soft tissues of a residual limb of a transtibial amputee. *Journal of Biomechanics*, 41, 1,897–1,909.

Radcliffe, C.W. & Foort, J. (1961). *The patellar tendon bearing below-knee prosthesis*, University of California, Berkeley, CA.

Radcliff, C.W., Foort, J. & Inman, V. (1961). The patella-tendon bearing below-knee prosthesis. *Biomechanics laboratory report*, University of California at Berkeley, Berkeley, CA.

Rommers, G. M., Vos, L. D. W., Klein, L., Groothoff, J. W. & Eisma, W. H. (2000). A study of technical changes to lower limb prostheses after initial fitting. *Prosthetics and Orthotics International*, 24, 28–38.

Ron Seymour (2002). *Prosthetics and orthotics: lower limb and spine*, Lippincott, Williams & Wilkins, Philadelphia, PA.

Sanders, J. E., Daly, C. H. & Burgess, E. M. (1992). Interface shear stresses during ambulation with a below-knee prosthetic limb. *Journal of Rehabilitation Research and Development*, 29, 1–8.

Sanders, J. E., Daly, C. H., & Burgess, E. M. (1993). Clinical measurement of normal and shear stresses on a trans-tibial stump: characteristics of wave-form shapes during walking. *Prosthetics and Orthotics International*, 17(1), 38–48.

Sewell, P., Noroozi, S., Vinney, J., Amali, R., & Andrews, S. (2012). Static and dynamic pressure prediction for prosthetic socket fitting assessment utilising an inverse problem approach. *Artificial Intelligence in Medicine*, 54(1), 29–41.

Sewell, P., Noroozi, S., Vinney, J., & Andrews, S. (2000). Developments in the trans-tibial prosthetic socket fitting process: A review of past and present research. *Prosthetics and Orthotics International*, 24(2), 97–107.

Silver-Thorn, M. B. (2003). Design of Artificial Limbs for Lower Extremity Amputees. In: The McGraw-Hill Companies, I. (ed.) *Standard Handbook of Biomedical Engineering & Design*. The McGraw-Hill, New York.

Steele, A. L. (1994). A survey of clinical CAD/CAM use. *JPO: Journal of Prosthetics and Orthotics*, 6, 42–47.

Stekelenburg, A., Gawlitta, D., Bader, D. L. & Oomens, C. W. (2008). Deep tissue injury: how deep is our understanding? *Archives of Physical Medicine and Rehabilitation*, 89, 1,410–1,413.

Tanner, J. & Berke, G. (2001). Radiographic comparison of vertical tibial translation using two types of suspensions on a transtibial prosthesis: a case study. *Prosthetics and Orthotics International*, 13, 14–16.

Tekscan, I. 2010. Tekscan F-scan pressure measurement system *F-scan Research*. 6.51 ed.

Tillander, J., Hagberg, K., Hagberg, L. & Branemark, R. (2010). Osseointegrated titanium implants for limb prostheses attachments: infectious complications. *Clinical Orthopaedics and Related Research*, 468, 2,781–2,788.

Vasconcelos, E. (1945). *Modern Methods of Amputation*, PhilosophicalLibrary, New York.

Zhang, M., & Roberts, C. (2000). Comparison of computational analysis with clinical measurement of stresses on below-knee residual limb in a prosthetic socket. *Medical Engineering & Physics*, 22(9), 607–612.

Ziegler-Graham, K., Mackenzie, E. J., Ephraim, P. L., Travison, T. G. & Brookmeyer, R. (2008). Estimating The Prevalence of Limb Loss in The United States: 2005 To 2050. *Archives of Physical Medicine and Rehabilitation*, 89, 422–429.

# 6 Wrist Movement with Ultrasonic Sensor and Servo Motor

## N A Abu Osman and N A Abd Razak

University of Malaya, Kuala Lumpur, Malaysia

## CONTENTS

## 6.1 COMPONENTS OF THE TRANSRADIAL PROSTHETICS MOTION SYSTEM

The prosthetics arm basically uses ultrasonic sensor to transfer any motion detection data to the microprocessor and microcontroller-based system as the input data. The ultrasonic sensor is one of the most accurate and reliable measurement tools to determine human motion intensity. An ultrasonic sensor uses transmitter and receiver

DOI: 10.1201/9781003196730-6

wave to get the reflection of any motion within 0–15cm. The sensor will be attached at the shoulder of the amputee as the replacement of tension cable in body-powered prosthetics (Stark and LeBlanc, 2004). Instead of using only motion detection, the patient does not have to worry about training their muscle system to operate the system as compared to the body-powered tension cable prosthetics.

The sensor that functions as the input will then generate the data into the microcontroller system that is placed inside the transradial part. This part of the transradial also consists of two servo motors that will operate as the replacement of motion of

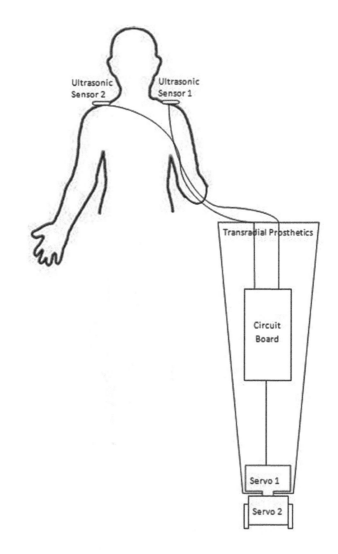

**FIGURE 6.1**   Structure of the new transradial prosthetics system.

the extension/flexion and supination/pronation (Figure 6.1). The servo motor also has its own limitation of rotation degree similar like the real transradial human movement. The servo motor can generate a maximum of 30Nm torque, which is much greater than the required energy to do daily tasks that usually only need around 10–30Nm. Servo 1 is used to generate the pronation/supination, while Servo 2 is used in flexion/extension. The power supply for the system comes from a 9V battery that is very light and long-lasting.

## 6.2 PHYSICAL SETUP

The procedure used to operate the transradial prosthetics is stated in Figure 6.2.

The electronic mechanism usually consists of three main parts that are the input, the controller, and the output systems (Figure 6.3). The electronic mechanism for this

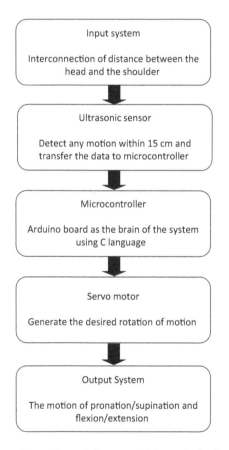

**FIGURE 6.2** The general procedure of the transradial prosthetics interconnection from the input to the output system.

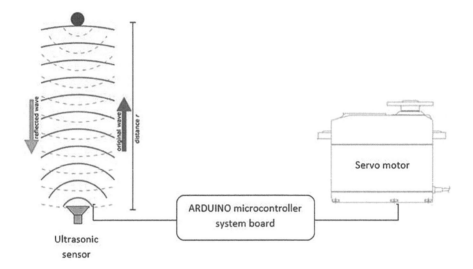

**FIGURE 6.3**   Flowchart of the input–output system.

transradial prosthetics uses the ultrasonic sensor as the input, the servo motor as the actuator output and the Arduino microcontroller system board as the brain of the system. Figure 6.3 shows the connection between each part based on the input–output system.

The ultrasonic sensor that functions as the input of the system generates the signal in analogue form. The data is then transmitted to the Arduino controller in digital form after entering the ADC converter. The data will then be recognized by the Arduino microcontroller system that uses the computer language to generate the memory of instruction. The data is then reconverted into the DAC converter to manipulate the servo motor motion as the output of the system. Note that the microcontroller can be programmed according to the need of the user using C programming language software. The ADC and DAC converters inside the microcontroller also play a major part in transferring the data and memory of the system.

## 6.3   CIRCUIT DIAGRAM

The main circuit is divided into four main parts (Figure 6.4). The input is from the ultrasonic sensor, the controller (Arduino microcontroller), the output that uses servo motor and the power supply to the system. The circuit combination of ultrasonic sensor uses pin 8 and pin 9 on the Arduino microcontroller board to receive the data input. The common servo output uses both pin 6 and pin 7 as in Figure 6.4. The circuit is generated by 9V and 5V power supplies that come either from the USB computer support or 9V battery. The details of the circuit components are reviewed in later sections of this chapter.

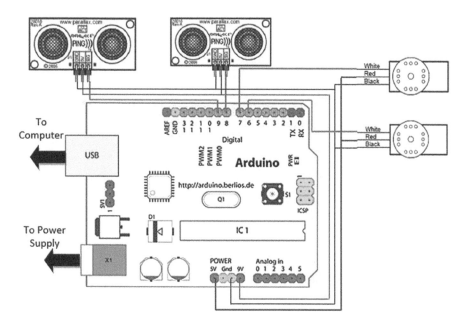

**FIGURE 6.4**   Main circuit of transradial prosthetics.

## 6.4   MICROCONTROLLER SYSTEM

From this block diagram (Figure 6.5), the plant of the system is divided into two parts: the actuator and the controller. The motor functions as the actuator or the device of the system. The motor rotation is used to control the speed and the desired direction of the transradial segment. This control plant also needs to determine first accordingly with the feedback of the system. Here, the feedback of the system works according to the sensor. If the sensor detects obstacles, the motor will rotate synchronously.

The microcontroller functions as the brain of the transradial movement system. The Arduino microcontroller circuit serves as the main controller of the robot system, controlling the system's main input and output. The microcontroller was chosen because of its ability to evaluate any sort of output/input device, whether analogue or digital. It also has 14 digital I/O pins (six of which can be used as PWM outputs), six analogue inputs, a 16 MHz crystal oscillator, a USB link, a power jack, an ICSP header and a reset button. The programme is stored in ROM (read only memory) and does not usually change. A microcontroller often receives feedback from the computer and controls it by transmitting signals to various components inside the device. It comes with everything you need to help the microcontroller; simply attach it to a computer through USB or plug it into an AC-to-DC adapter or battery to get started. The benefit of using an Arduino microcontroller is that the program memory is built into the chip, and it is limited in size due to its onboard memory. For research, the Arduino microcontroller can be moved to a real circuit board.

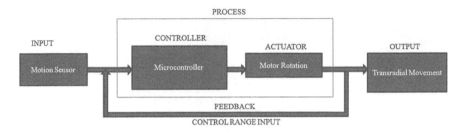

**FIGURE 6.5** Block diagram of the system.

In this project, 4 pins in the Arduino microcontroller board were used, which were pins 6–9. Pins 6 and 7 were used to generate and transfer the data to servo motor, while pins 8 and 9 functioned as the input of the system to collect the data from the ultrasonic sensor.

## 6.5 INPUT–OUTPUT SYSTEM

### 6.5.1 ULTRASONIC SENSOR

The ultrasonic sensor is used for this transradial system to determine the range of detection and to give the input to the motor controller system. The ultrasonic sensor generates a wave of detection about 10–15cm obstacle. If the Arduino microcontroller receives the input pulse, it will convert the data and read by the memory of the system to generate the servo motor.

The reason for choosing the ultrasonic sensor and not others such as IR is that the ultrasonic sensor is more accurate in reading the receiver value. For example, it reads an integer value and can be directly understood by HMI (human machine interface) rather than in binary if an IR sensor is used. Also, the ultrasonic sensor does not reflect because of the types of material or the colour of the obstacle. It will only reflect other types of wave or frequency. For example, it will work poorly if it is too close to others. For this transradial system, we were assigned to use pins 8 and 9 as the receiver of the input signal to the microcontroller board system.

### 6.5.2 ULTRASONIC RANGE DETECTION SENSOR

The ultrasonic sensor wave has been programmed to detect any motion within the range of 0–15cm and is placed on the shoulder. The motion of head to the left and right sides or the upside and down sides of the shoulder will give the detection to the ultrasonic sensor. Figure 6.6 shows how the motion of the head and the shoulder gives the detection to the wave of ultrasonic sensor. Generally, the ultrasonic sensor detects about 200–300cm range if the user is inside a building or a room (Figure 6.7). The range will be reduced immediately after the head blocks the wave signal and will be continued until the range is about 15cm.

Figure 6.7 shows the graph of the ultrasonic detection range. The sensor was worn by the amputee by standing inside a room. That is why the detection range is about

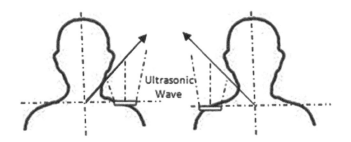

**FIGURE 6.6**   Ultrasonic motion is required to give the input for detection.

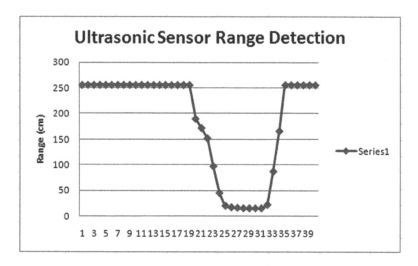

**FIGURE 6.7**   Ultrasonic sensor detection range.

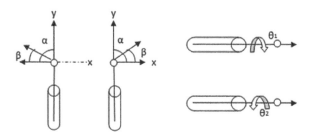

**FIGURE 6.8**   Kinematic model of transradial.

250cm since the height of the room was about 300–400cm including the height of the amputee. The graph then shows the immediate decrease to 15cm (the range of the input for the microcontroller). The amputee then will have to do the required motion to give the detection signal to the system (Figure 6.8). The graph later increases slightly to the normal position of the head and shoulder of the amputee.

### 6.5.3 SERVO ACTUATOR

Servo motor is used as the output of the transradial prosthetics system. There are several types of motor used to design a robotic system. Some common examples are the servo motor, stepper motor and DC motor. Each has its own classifications, advantages and disadvantages, but the most common motor used in robotic systems is usually the servo motor. Besides having high torque, servo motors have a capability to rotate precisely according to the degree assigned. For transradial prosthetics, the capability of the servo to generate the motion is required. The chosen maximum torque that can be applied by the servo is 13Nm. This is because in designing the robotic transradial, it needs to deal with the high-rate motion such as pick-and-place motion and rotation motion.

Servo motors also give more precise rotation. A DC motor may have faster and more accurate degrees of rotation, but the torque generated is not usable to generate the motion. Stepper motors, on the other hand, give accurate degrees of rotation, but move step by step, and thus are not suitable for transradial movements that require smooth and sharp degrees of rotation. The servos give both high torque and precise rotation.

Note that the servo motor's degree of rotation is about 180°, which is suitable for supination/pronation and flexion/extension movements. The weight is about 0.057kg and the dimension of the servo motor is suitable to be a part of the transradial system based on the anthropometry theoretical. The voltage to generate the motion is only about 4.8–6V. The voltage supply needs to be attached apart from the system since it needs of a lot of voltage to run the motor.

### 6.5.4 ROBOTIC ACTUATOR PRINCIPLE

The arm consists of the transradial part (elbow disarticulation and wrist disarticulation) until the arm. The focus in this study is on how the transradial part robotic arm movement will be suitable with the human body part that is going to be replaced. The human body moment of inertia needs to be considered as to make the rotation of the arm axis even that the movement of the pronation/supination is not as greater as the flexion/extension movement.

Mechanical design for each part and joint of the arm will be based on the high-torque servo motor. This actuator will become the major part in holding and rotate the servo motor movement for the pronation and supination.

Two mechanisms will be used in this study. First, the rotation of the first transradial part used a servo motor to make the pronation ($\theta_1$) and supination ($\theta_2$) movements. It needs to consider the clockwise and the anticlockwise rotations of the servo motor. The rotations of $\theta_1$ and $\theta_2$ depend on the input data from the ultrasonic sensor. By taking x-axis as the origin for the degree of rotation, $\theta_1$ and $\theta_2$ might not be equalized as compared to the real human hand. The degree might be greater or lesser but should be 85–90° for $\theta 1$ and $\theta_2$.

The second mechanism is on the flexion ($\alpha_1$ and $\beta_1$) and extension ($\alpha_2$ and $\beta_2$) that also use the servo motor with the joint. They also have clockwise and anticlockwise rotations. The extension's degree of rotation for a human wrist is 70–90°. As y-axis is the origin, Figure 6.7 shows how the maximum degree of extension can be reached by 90° ($\alpha_2$) that depends on the force that is applied to that part. But usually the human hand movement will only reach up to 70° ($\beta_2$) in daily tasks. The flexion's maximum degree of rotation is usually between 80° ($\beta_1$) until 90° ($\alpha_1$) depending on the force applied. But, in daily tasks, human hand uses flexion more than extension. This causes a higher degree of movement in flexion.

The two mechanisms of the servo motor that are used to generate the transradial motion consist of the most relevant size and power needed for the transradial movement. The mechanisms use the principle of robotics that is like a part in the motion of articulated robotics mechanism. The general robotics term for the transradial motion is known as 2-rotational (2R) robotic movements. That is because of the transradial motion involves two rotational movements or known as two degrees of freedom. Servo 1 is considered as joint 1 in Figure 6.9 is where the pronation and supination movements take place. The rotation has clockwise and anticlockwise movements with relevant degree and it depends on the x–y–z axis. Servo 2 uses the same procedure to generate the flexion and extension movements.

Figure 6.10 shows the parameters of the transradial robotics to determine the transformation of each joint. Note that $Z_0$ represents the first revolute joint. $X_0$ is parallel to the reference origin of x-axis for the purpose of convenience and as the

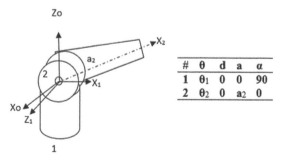

**FIGURE 6.9**    The prototype of the transradial segment with the servo motor.

| # | θ | d | a | α |
|---|-----|---|-----|----|
| 1 | $\theta_1$ | 0 | 0 | 90 |
| 2 | $\theta_2$ | 0 | $a_2$ | 0 |

**FIGURE 6.10**    Parameters for the transradial robotics.

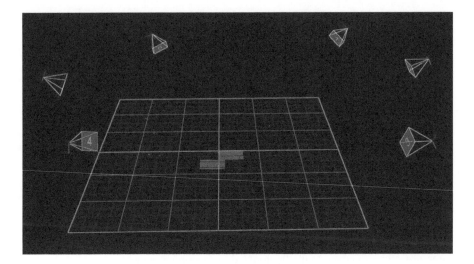

**FIGURE 6.11**  A bird's-eye view of the cameras and force plates setup. The six cameras were placed at the four corners of the room and two the force plates. The two force plates were embedded in the middle of the capture volume.

base of the robotic system that does not move. $Z_0$–$X_0$ gives the first joint movement even when it does not move. $Z_1$ is assigned at joint 2. $X_1$ will be normal to $Z_0$ and $Z_1$ because these two axes are intersecting. In $Z_0$–$X_0$ there will be a rotation of $\theta_1$ to bring $X_0$ to $X_1$. Then, both $X_1$ and $Z_1$ are assigned as zero to align the Z and X together in order to rotate $\alpha$ and bring $Z_0$ to $Z_1$ by taking right-hand rule as the reference. Then, $Z_1$ rotates with an angle of $\theta_2$ to bring $X_1$ to $X_2$ and move along the $a_2$ to bring the origins together (Sensinger & Weir, 2008).

From the data, the $4 \times 4$ matrix was generated using the same articulated robotic mechanism (Figure 6.11). The development focuses only on two joints, which are referred to as joint 1 and joint 2. The motion transformation of each two successive joints can be written by into the A matrix for each joint. The motion transformation for the transradial segment depends on the degree of rotation ($\theta$), distance of the joint to be rotate (d), the distance of the joint (a), and the degree of rotation from rotation to parallel. In the transradial robotic principle, we do not need to consider the parallel motion since both servos give the rotational movement.

To measure the real degree of rotation that the transradial should have, a few values have been calculated by comparing them with the daily transradial activities. The degree of the A1 joint is usually between 70–85° and the maximum is up to 90° and that is followed by the robotic principle for joint 1. The transformation for A2 joint also uses several values that depend on the daily tasks during flexion and extension. The transformations for joint 1 and joint 2 give the relevant theoretical static values. To get the dynamic experimental value, an experiment of motion analysis of the transradial motion was done by applying the transradial motion and analyzed by using the motion analysis Vicon data.

### 6.5.5 POWER SUPPLY

Since the Arduino board is used, the advantage is that it is already attached to a port of the shares the voltage with the device that is attached to it. Besides, the board also provides the multifunctional USB port that can easily generate a power supply after being attached to the computer. The power source is selected automatically. External (non-USB) power can come either from an AC-to-DC adapter (wall wart) or battery. The adapter can be connected by plugging a 2.1mm centre-positive plug into the board's power jack. Leads from a battery can be inserted in the Gnd and Vin pin headers of the power connector.

For the system, the board generates 9V and 5V (Table 6.1). The voltage supply for the system and the servo motor needs to be separated from each system to give a full power to the motor to generate high torque. 5V batteries are used to run the servo motor. The reason for separating the 5V power supply from the driver and the motor is that the servo motor needs its own power to run the motor. If the voltage supply to the motor is very low or the same as the others, it may cause the motor to be too weak to run.

The specifications of the power supply divider where the 5V batteries are for both servo motors while the 9V batteries are for the controller and the ultrasonic sensor. The board can operate by an external supply of 6–20 V. If supplied with less than 7 V, the 5V pin may supply less than 5V and the board may be unstable. If more than 12V is used, the voltage regulator may overheat, and this can damage the board.

### 6.5.6 MOTION ANALYSIS

The mechanisms of transradial prosthetics movement, which are the pronation/supination and flexion/extension, were tested using the motion analysis. The experiment setup uses the combination of six MX-F20 infrared cameras into the Vicon Nexus 6.1.109 that makes up the capture system used in this study. The infrared cameras are used to investigate the motion done by the amputee user. The six MX-F20 cameras were positioned at each corner of the room and the midpoint of the room's width. The arrangement of the connection of all the cameras gives the full dimensions of the room, which has the maximum volume of 37.5m³. Afterwards, only object within this area may be detected by the motion signal. With the 1,600 × 1,280 pixels of all cameras, any motion or object within the area can be tracked in real time. The MX-F20

**TABLE 6.1**
**Power Supply Distribution**

| Voltage Power Supply | User |
| --- | --- |
| 9V | Ultrasonic Sensor 1 |
| 9V | Ultrasonic Sensor 2 |
| 5V | Servo Motor 1 |
| 5V | Servo Motor 2 |
| 9V | Arduino Microcontroller Board |

cameras were synchronized via the MX-Ultranet box that is used to transfer the data of both analogue and digital data to the host computer.

The Vicon Nexus was used to provide a complete calculation and dimensions of the space for the calibration of the MX-F20. Any time the subject turns, it provides an angle to the cameras, allowing them to catch the motion. Based on the tuning, the Vicon programme helps the motion to recreate a 3D model in space. During the experiments, the system's calibration must be calibrated per time. Despite the fact that the cameras are securely fixed on the wall, the area of the room can be disrupted from time to time, and the identification of the cameras may change. Before the trials began, static and dynamic calibrations were performed. The object must stand in the centre of the device for static calibration, while the object must switch from one location to another for dynamic calibration. This technique is typically used to ensure that all cameras are operational, and that the data transferred is accurate.

The detection of any motion within the area of the cameras depends on the reflection of the markers. Markers are spheres that reflect the light from the strobe to the camera. The dimensions are about 14mm in diameter and there are 32 markers positioned all around the body. The positions of the markers and the description of each point is shown in the appendix. During the experiment, the markers are placed on both normal and amputation subjects. To have a full view of movement, the markers are put all around the body instead of on transradial part only. To have accurate results, the subject was advised to wear tight-fitting clothes to prevent artefacts from the movement of loose clothing. This is because the camera captures any movement of the markers.

Figure 6.12 shows the marker positions for both real time and of the Vicon software. Note that the A-C shows how the measurements are taken for each body segment so that the markers are put at the required positions. All 32 markers that are detected by the camera will be identified by the Vicon system to construct the body motion segment as in D in Figure 6.12. The colours represent the left and right sides of the body.

## 6.6 MECHANICAL RESULTS OF THE DESIGN

### 6.6.1 MOTOR ROTATION

The most critical problem occurred to determine the correct angle for the rotation of the servo motor. As explained in the previous chapter, the chosen rotation of the servo motor for both transradial motions was 0–180°. However, the test was done before the installation of the servo motor to the transradial prosthetics. In order to have the maximum rotation, the power supply was set up to 5V. The trials were divided into two ranges of rotation, which were from 0–90°and 0–180°. The results are shown in Table 6.2. For the 0–90° range of rotation, the servo motor gave precise value, but it did not achieve exactly the 90° of rotation. The result achieved was only about 87–89° of rotation. For the maximum degree of rotation, which was 180°, the result was similar. The servo motor achieved only up to 177–179° of rotation. The inertia of the servo motor played a major role to determine the rotation of the motor. Frequently, when the motor rotated, there would be a delay due to the inertia before

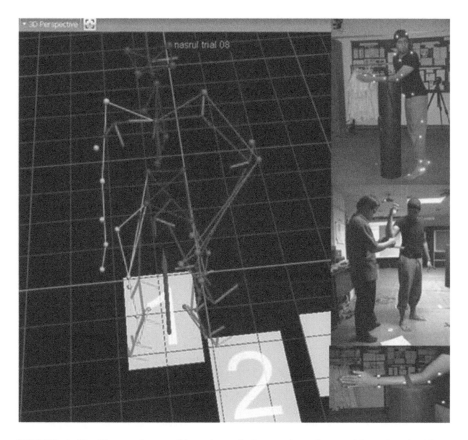

**FIGURE 6.12** The marker positions in real time and marker positions by the Vicon software.

**TABLE 6.2**
**Rotation Trials**

| Trial | Rotation 0–180° | Rotation 0–90° |
|-------|-----------------|----------------|
| 1 | 179° | 88° |
| 2 | 179° | 87° |
| 3 | 178° | 87° |
| 4 | 177° | 87° |
| 5 | 177° | 85° |

the motor moved. The other reason was the lack of power dispersed before being used to generate the motor. However, neither problem interrupted the motion system since the maximum human daily tasks that use the transradial part is only about 70–85°, and this was already achieved.

### 6.6.2 Force of the Servo Motor

Other criteria investigated was the general torque applied to the servo motor. Based on the criteria of the motor, it was suggested that the servo motor could handle a weight of up to 13Nm, which is about 130N of the desired force to be applied. But for the usual tasks in daily life, the force needed was around only 20–30N. To open a heavy mounted door, the desired force is about 30N. So, the experimental setup was applied to the servo motor by applying the desired force to the servo motor (Figure 6.13). Based on the experiment, it occurred that the maximum force that can be generated by the servo motor was 2.5kg, which is equivalent to around 24–26N (Figure 6.14). The power of the servo motor could achieve higher values if the voltage applied is maximized to 6V. But the idea of using the highest torque to replace the transradial motion was good enough since the force and torque can give similar characteristics to the real hand activities. Note that, for transradial prosthetics, both servo motors had the same characteristics and performed similar rotations.

(a)                    (b)                    (c)

**FIGURE 6.13** (a) Shows the initial position of the servo motor, (b) shows where the motor is rotated to 90° vertical, but the result gave about α (85–88°), (c) shows the β (85–88°) although the theoretical value is 90° of rotation.

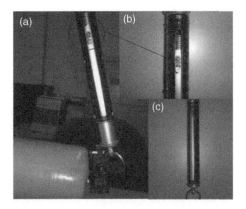

**FIGURE 6.14** (a) Shows the weight scale used to measure the weight of the pulling force by the servo motor as scale view in (b) while (c) shows the weight scale.

## 6.7   THE ELECTRONIC MECHANISM

### 6.7.1   EFFICIENCY OF ULTRASONIC SENSOR

The difference in this project is the use of the new technique in moving the transradial prostheses by using ultrasonic sensor as the input system. The experiment first involved the installation of the transradial prosthetics to the amputee. The ultrasonic sensor was placed at the top of the shoulder. The sensor was placed vertically parallel to the amputee's head. It was very sensitive to any motion, especially to solid material and not reflective materials. Based on the ten trials done to the sensor, the sensor's range of detection was about 0–16cm. But in this project, the head of the amputee was the object required to be reflected. So, the surface and the shape of the head were already known to be good reflectors and should not give problem to the input–output system. However, conditions like wet or oily hair may interrupt the range of the motion detection.

Based on the result of the experiment (Table 6.3) it was found that the strongest detection was within the value of about 15cm or about 45–55° bend of the head to shoulder. The surface size of the head and shoulder did not really matter but the circuit depended more on the range of detection.

As a result, the range can be determined from these three different graphs of signal wave. Figure 6.15(a) shows how high detection and the pins 8 9 of the microcontroller sent the data to rotate the servo motor.

Figure 6.15(b) shows low detection. So, the input was negligible and no data were transferred to the microcontroller. Figure 6.15(c) shows how the ultrasonic sensor worked with the detection of motion. First, it detected the head position and sent the input data to the microcontroller when the servo motor rotated. The second wave shows how the second range of detection has occurred and the second rotation motor will follow accordingly.

After several tests were done on the ultrasonic input system, it was found that the first three trials were not efficient in detecting any motions. After some problems on the circuit had been fixed, the input system managed to detect the presence of motion within its range. Ultrasonic sensor was chosen for the motion detection because the range was read in integer, and it was not affected by the types and colour of material.

## TABLE 6.3
### Condition of the Detection Wave at Each Point

| Trials | Distance Between Shoulder and Head | Condition |
|---|---|---|
| 1 | 17.50cm | Too far, not detected |
| 2 | 17.00cm | Far, not detected |
| 3 | 16.50cm | Slow wave, depending on the head condition |
| 4 | 16.00cm | Slow wave, but not enough for input system |
| 5 | 15.50cm | Wave detected |
| 6 | 15.00cm | Strong wave detected |
| 7 | 14.50cm | Strong wave detected |
| 8 | 14.00cm | Strong wave detected |
| 9 | 13.50cm | Very strong wave detected |
| 10 | 13.00cm | Maximum wave detected |

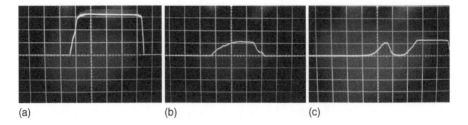

(a)                              (b)                              (c)

**FIGURE 6.15**    (a) High detection (b) Low detection (c) Both detections.

However, a problem occurred with the power supply where the system shared the same power supply with Arduino microcontroller. That was the why reason the system could not operate if the voltage supply were less than 8V.

### 6.7.2  BATTERY PERFORMANCE

The transradial prosthetics was worn by the patient for about an hour, and the battery performance was tested every ten minutes. Based on the results shown in Table 6.4, the motor and the controller could run continuously for up to 60 minutes. As recorded, the voltage of the microcontroller board and ultrasonic sensor decreased from 9.05V to 8.12V in one hour. For the servo motor power supply, the 6V slightly decreased to 5.12V within one hour. The two servo motors used a different power supply from the board to generate more voltage to give better rotation.

For the 9V supply, the decrease of the battery performance was normal and did not interrupt the system. But to get the best performance of the system, the battery may need to be changed regularly. The battery performance might not interrupt the microcontroller system. But the performance and the sensitivity of the ultrasonic sensor may be interrupted. Even though the ultrasonic controller was able to transmit the wave, the receiver wave may not work properly, and thus gave zero feedback to the system. The battery could be changed to a longer lasting battery such as alkaline, or recharged from time to time. The body-powered prosthetics and electric-powered prosthetics need to be taken off from time to time by the patient anyway.

For the servo motor supply, the 6V power supply was used rather than the required 5V power supply (Table 6.5). This was due to the 1.5V × 4 of battery type that gave

**TABLE 6.4**
**9V Battery Performance in One Hour**

| Time (min) | Voltage (9V) |
| --- | --- |
| 10 | 9.05 |
| 20 | 8.92 |
| 30 | 8.75 |
| 40 | 8.54 |
| 50 | 8.32 |
| 60 | 8.12 |

**TABLE 6.5**
**6V Battery Performances in One Hour**

| Time (min) | Voltage (6V) |
|---|---|
| 10 | 5.95 |
| 20 | 5.72 |
| 30 | 5.65 |
| 40 | 5.32 |
| 50 | 5.26 |
| 60 | 5.12 |

6V. The battery performance decreased faster after being used by the servo motor for rotation. The voltage supply was the key factor to rotate the servo motor with the stable speed and accurate angle of rotation.

## 6.8 MOTION ANALYSIS

### 6.8.1 MOTION ANALYSIS COMPARISON

The motion analysis comparison of the angle of rotation is the main interest in this study. Eight cameras of the Vicon motion analysis system were used to collect and analyse the movement data from one transradial prosthetics user. Thirty-nine spherical reflective markers were placed from the lower limb until the shoulder, transhumeral and transradial. The markers were placed in the rigid body segments and the parameters of each segment were also collected, namely the shoulder depth, elbow width, wrist circumference and hand thickness. The kinematic model that covered the upper limb and lower limb was created using the Vicon bodybuilder software and the upper limb joint angles were calculated (Stephanie et al., 2008) (Figure 6.16).

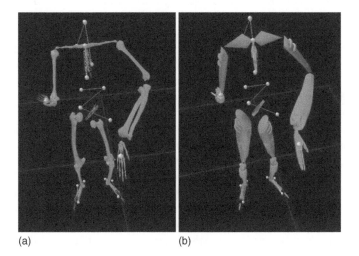

(a)                                    (b)

**FIGURE 6.16** (a) shows the skeleton system where the markers were attached including the transradial part using Vicon system, while (b) shows the mechanism in the robotic system.

**TABLE 6.6**

**During Each Task, the Max, Min, and Range of Motion (in Degrees)**

| Type of Motion | Range of Motion (ROM): Biological Hand | | | | | | | Range of Motion (ROM): Transradial Prosthetics | | | | | | |
| | Test 1 | Test 2 | Test 3 | Test 4 | Test 5 | SD | Average (ROM) | Test 1 | Test 2 | Test 3 | Test 4 | Test 5 | SD | Average (ROM) |
|---|---|---|---|---|---|---|---|---|---|---|---|---|---|---|
| Flexion | 21 | 21.5 | 20.3 | 21.9 | 20.2 | 0.74 | 20.6 | 18.7 | 22 | 24 | 17 | 19.2 | 2.27 | 22.4 |
| Extension | 55 | 57 | 57 | 56 | 57 | 0.89 | 57.3 | 40 | 39 | 41 | 41 | 40 | 0.84 | 41 |
| Pronation | 55 | 54.6 | 55.2 | 55.7 | 55.4 | 0.41 | 55.7 | 50 | 49.8 | 50.1 | 50 | 50 | 0.11 | 50.4 |
| Supination | 50.2 | 47.2 | 48.4 | 51.2 | 50.2 | 1.61 | 50 | 89.7 | 88.7 | 88.9 | 89.1 | 89.8 | 0.49 | 89.3 |

The subject completed four simulated transradial general movements, which are pronation/supination and flexion/extension. Each movement consisted of five trials. The best three were chosen, and the results are shown in Table 6.6. For the extension and flexion, the subject was asked to use the transradial prosthetics and moved it from the initial position to the final position. The maximum and minimum results were based on the angle of rotation. The subject repeated the task five times with the transradial prosthetics, and they were compared with the biological hand movement. Basically, the subject was asked to do each task separately, such as moving the flexion from initial position until the maximum position, moving the extension from initial position to final position, moving the supination from the initial position to final position, and moving the pronation from the initial position to final position. The results are shown in Table 6.6 and Table 6.7.

The averages of all trials for each task are shown in Table 6.6. The table simplifies the results by taking the appropriate values, and the averages of each motion for the transradial movement are specified in maximum, minimum and range. An issue arose with the wrist marker since the area was only filled by one marker and the degree of rotation could affect reliability. The range of natural hand motion during the flexion movement was 20.7° on average, while the prosthetics showed around 22.9° on average. The maximum value of hand flexion is typically 85–90°, but the value varies depending on how the muscle is stretched to achieve that position. The extension motions of the natural and prosthetic hands were approximately 57° and 41°, respectively. After some experiments, the transradial prosthetics offered a lower value due to the reduced capability of the servo motor, but the angle was still sufficient to perform everyday tasks that require extension motion. This two extension and flexion movements demonstrated that the conditions for performing everyday activities such as opening a door and filling a cup can be met.

For the pronation movement, the range of rotation of the prosthetic hand was about 55.7° (Stephanie et al., 2008). That was almost the same with the normal hand that showed 50.4° of rotation. The pronation movement for the daily tasks is usually between 85°–90°. Even though the required range is higher, the degree of rotation between the normal hand and the prosthetic hand was quite like each other. The supination movement usually only takes about 50–55° (Stephanie et al., 2008), which is

## TABLE 6.7
## Input, Output and Motion Analysis of Transradial Prosthetics System

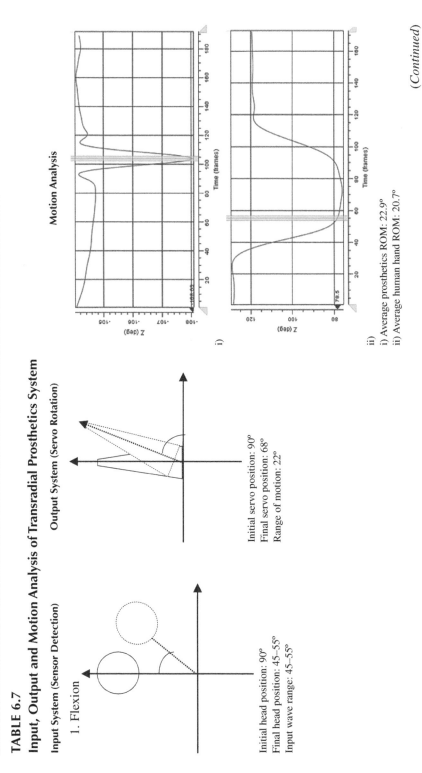

**Input System (Sensor Detection)**

1. Flexion

Initial head position: 90°
Final head position: 45–55°
Input wave range: 45–55°

**Output System (Servo Rotation)**

Initial servo position: 90°
Final servo position: 68°
Range of motion: 22°

**Motion Analysis**

i)

ii)

i) Average prosthetics ROM: 22.9°
ii) Average human hand ROM: 20.7°

*(Continued)*

**TABLE 6.7 (Continued)**

| Input System (Sensor Detection) | Output System (Servo Rotation) | Motion Analysis |
|---|---|---|

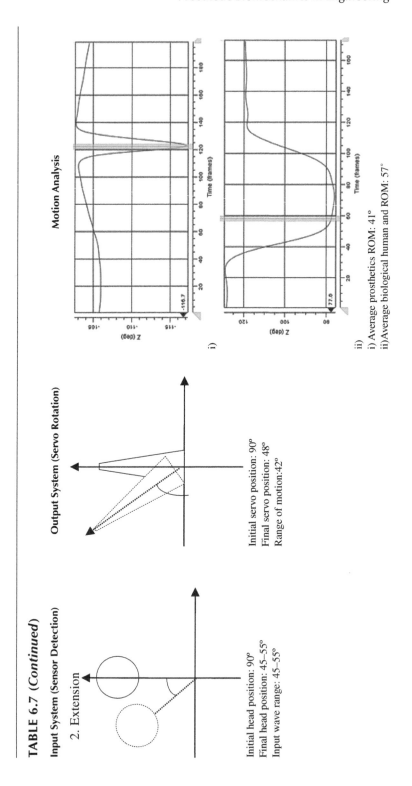

**Input System (Sensor Detection)**

2. Extension

Initial head position: 90°
Final head position: 45–55°
Input wave range: 45–55°

**Output System (Servo Rotation)**

Initial servo position: 90°
Final servo position: 48°
Range of motion:42°

**Motion Analysis**

i)

ii)

i) Average prosthetics ROM: 41°
ii)Average biological human and ROM: 57°

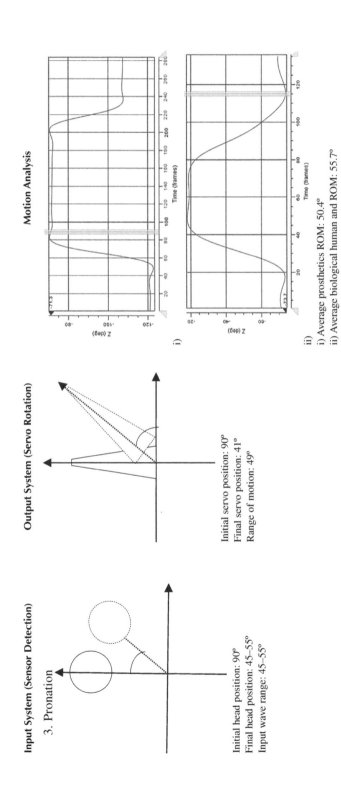

**Input System (Sensor Detection)**

3. Pronation

Initial head position: 90°
Final head position: 45–55°
Input wave range: 45–55°

**Output System (Servo Rotation)**

Initial servo position: 90°
Final servo position: 41°
Range of motion: 49°

**Motion Analysis**

i)
ii)
i) Average prosthetics ROM: 50.4°
ii) Average biological human and ROM: 55.7°

*(Continued)*

**TABLE 6.7 (*Continued*)**

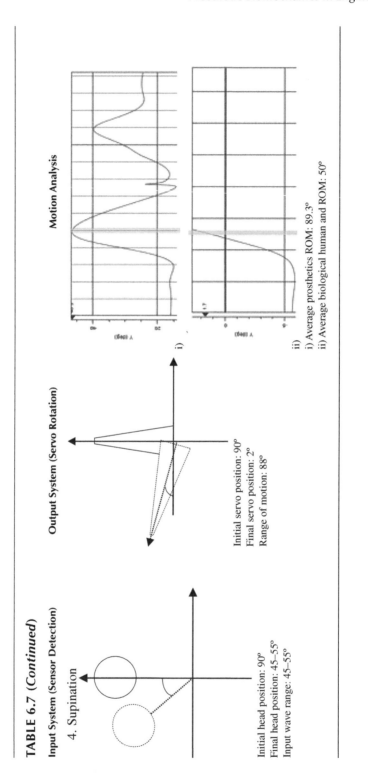

Input System (Sensor Detection)

4. Supination

Initial head position: 90°
Final head position: 45–55°
Input wave range: 45–55°

Output System (Servo Rotation)

Initial servo position: 90°
Final servo position: 2°
Range of motion: 88°

Motion Analysis

i)

ii)

i) Average prosthetics ROM: 89.3°
ii) Average biological human and ROM: 50°

similar to the motion for normal hand that gave about 50° of rotation. But the transradial prosthetics showed the maximum level of 89.3° of rotation.

Table 6.7 depicts the device from the input to the output of the motion analysis. The first column describes the input device, which used head motion to assign the input wave to the ultrasonic sensor. The angle of rotation needed for the head to bend to provide the maximum input wave was approximately 45–55°, or 14–16cm. The second column is concerned with the angle of rotation of the servo motor in order to produce motion. The final column discusses rotation of motion using motion analysis.

Since the subject did not overstress his muscles to achieve the optimum angle of rotation, the regular hand had a general rotation to perform everyday tasks. However, the values obtained from the Vicon motion study is accurate for performing basic tasks in everyday life. The prosthetic hand, on the other hand, had a greater angle of rotation to perform everyday activities but fell short of the optimum angle of rotation. Any results indicated a lower angle of rotation. This was attributed to a shortage of power supply following a number of experiments. That was also attributed to the spinning of the servo motor, which had its own inertia to produce motion. The angle of rotation was then adjusted as a result of the microcontroller's programming scheme and the motor's capacity. However, the aim of creating a transradial prosthetic hand that performs similarly to a human hand was met.

Based on Table 6.7, almost all the results gave a similar sinusoidal graph, but in terms of robotics principle, the angle of each rotation can be maximized up to 90°. The 2R (2-rotational) robotics theoretically can achieve up to 90°. But the results only gave the required rotation. Based on the graph in Table 6.7, most of the motions done by the robotic servo motor displayed a sharped slope than the motion done by human hand. This was due to the characteristics and speed mechanism of the motor. Human hand gave smooth and dynamic motion but the transradial prosthetics gave a direct synchronous motion. That is why the slope of each motion was slightly different but at the end, the results showed similar position. This is not a problem to the system since the speed and characteristics are greater than that of the human hand.

Figure 6.17 shows the two motions in one system. This was for further study to encourage the motion to determine a few tasks. For examples, to open a door, to fill in a cup, to drive a car etc. The results showed that both motions give similar motion production like human hand to generate the daily task motions such as combination of both flexion and supination.

## 6.9  CONCLUSION

The main objectives of this study are to develop a new approach for generating transradial prosthetic movement using ultrasonic sensor, to measure the performance of the results based on the kinetics and kinematics principle, and to compare them with the current prosthetics and real hand motion. The parameters used to measure the transradial prosthetics include the residual limb interface pressure, angle of motion analysis and the control system mechanism. The rotation was found to be lower than the real hand motion. The analysis of kinetic and kinematic parameters indicated that the rotation of the transradial prosthetics achieved the criteria for a daily tasks motion.

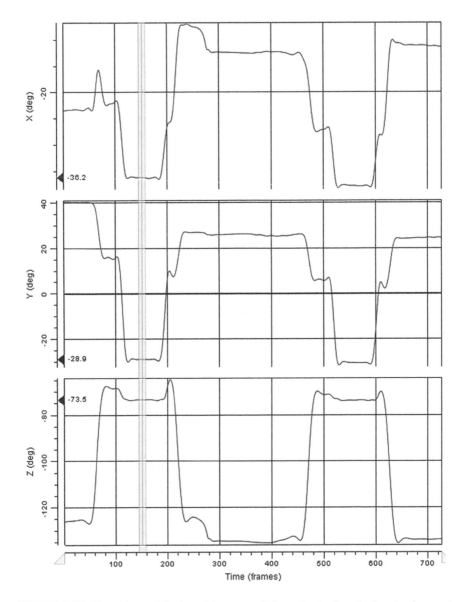

**FIGURE 6.17**  Pronation and flexion of the transradial prosthetics for a further development such as opening a door and driving a car.

The subjects were comfortable with the transradial prosthetics and showed confidence with the transradial prosthetics designed. This was acknowledged verbally and observed in its ability to perform with the minimal significant difference in movement parameters when compared to the real hand motion. The pressure analysis showed that the pressure at the residual limb was lesser than the body-powered prosthetics.

The limitation of the study was the small number of subjects. As mentioned before, the number of amputees who registered for the upper limb prosthetics was very limited due to the small number of the device. Future work should expand the motion system to give more motion to each upper limb. This includes the trans-humeral and the hand that are used in the daily tasks.

In conclusion, this study offers an interesting finding about the upper limb pros-thetics that uses a new mechanism in the transradial motion. Such a small develop-ment can make a big contribution to other research, especially in prosthetics and orthotics.

## REFERENCES

Sensinger, J.W. & Weir, R.F. (2008).User-modulated impedance control of a prosthetic elbow in unconstrained, perturbed motion. *IEEE Transactions on Biomedical Engineering*, vol. 55, p. 3.

Stark, G. & LeBlanc, M. (2004). Overview of body-powered upper extremity prostheses, in *Functional restoration of adults and children with upper extremity amputation*, Demos Medical Publishing, New York, vol. 7, p 29.

Stephanie L. C., Highsmith, M. J., Maitland, M. E., & Dubey, R. V. (2008). Compensatory movements of transradial prosthesis users during common tasks, *Clinical Biomechanics*, *23*, 1128–1135.

# 7 Study of a Dermo and Seal-In X5 Liner

## N A Abu Osman

University of Malaya, Kuala Lumpur, Malaysia

## S Ali

Oslo Metropolitan University, Oslo, Norway

## CONTENTS

DOI: 10.1201/9781003196730-7

## 7.1  INTRODUCTION

Transtibial amputation is the most common form of lower limb amputation. Surgeons aim to balance three conditions to achieve an effective amputation and a good residual limb: correct nerve ending positioning, proper bone thickness and sufficient soft tissue padding at the residual limb end. A normal amputation is performed when 20–50% of the tibia length is retained and at least 8cm of tibia length below the knee joint is preserved to allow for a good prosthetic fit. One surgical procedure involves extending the posterior flap and bringing the superficial posterior leg muscles, gastrocnemius and soleus, forward over the end of the residual limb to provide padding for the protruding distal end of the tibia and fibula.

Shear force, moisture, weight distribution, and temperature are all factors that affect the use of prosthesis and the interface between prosthetic socket and skin (Bui et al., 2009). The key challenge for prosthetic recovery is the inability to use and accept prosthesis due to discomfort inside the prosthetic socket (Chadderton, 1978;). The mechanical contact between the residual limb and the socket may influence the comfort and usability of the prosthesis. To prevent skin problems and discomfort during everyday activities, extra care should be taken during the design and fitting of the socket (Zhang and Roberts, 2000).

During their everyday activities, transtibial prosthesis (TTP) users encounter various paths such as level ground, ramps, stairs and other uneven surfaces. The ability to navigate environmental hazards such as bridges, rocky terrain, and stairs is an important factor in functional independence. Because of the loss of the foot and ankle system and the observed high interface strain, studies indicate that the lower limb amputee is significantly impacted when coping with environmental obstacles such as slopes and stairs.

Over the last few years, there has been an increasing awareness of the importance of assessing and calculating prosthetic and orthotic practices (Hoxie, 1995). There is a need for reliable and valid self-report instruments that can assist facilities in evaluating patient outcomes. Many questionnaires for prosthetics and orthoses have been created by researchers to assess patient satisfaction and issues with prostheses and orthoses.

### 7.1.1  Pelite or Polyethylene Foam Liner

Pelite or polyethylene foam is available in a variety of thicknesses and durometers (hardness). After heating, pelite is thermoformable and can be built over the positive cast. The benefit of pelite and other similar materials is that they are easily adjustable and can theoretically be used for the supracondylar wedge of the prosthesis. As the volume of the residual limb varies, additional pelite may be glued to the liner.

### 7.1.2 SILICONE LINERS

Silicone liners form a strong bond between the residual limb and the socket, providing a better interface and suspension than other socket forms. Silicone liners are often said to provide good skin safety and to minimize friction between the residual limb and the socket (Figures 7.1, 7.2, 7.3).

## 7.2 FABRICATION OF PROSTHESIS

### 7.2.1 MARKING, MEASUREMENT, CASTINGS POP ON RESIDUAL LIMB

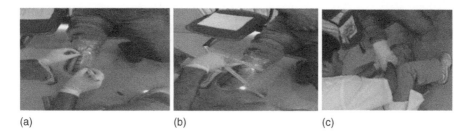

(a)  (b)  (c)

**FIGURE 7.1** (a) Residual limb marking. (b) Taking measurements from residual limb. (c) POP wrapping on residual limb.

### 7.2.2 FILLING AND MODIFICATION OF NEGATIVE CAST, MAKING THE TEST SOCKET

(a)  (b)  (c)

**FIGURE 7.2** (a) Filling of negative cast. (b) Modification of negative cast. (c) Making the test socket.

### 7.2.3  Assembly of the Prosthesis

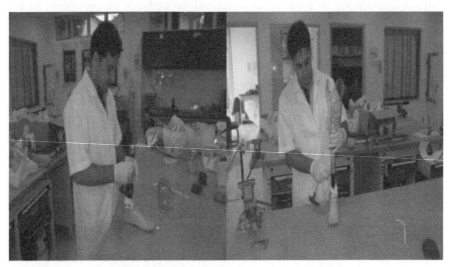

**FIGURE 7.3**   Check socket prosthesis assembly.

### 7.2.4   Making PVA Bags and Lamination

The first PVA bag is moulded on the plaster model for isolation during the lamination process. On the optimistic model, cotton stockinet and glass stockinet are rolled over. Carbon fibres are used by some amputees to strengthen the socket. The epoxy resin is poured into the second PVA bag and evenly applied to all surfaces of the positive model (Figure 7.4).

### 7.2.5   Alignment

During the complex process, the prosthetist watches the amputee's gait and listens to his or her input. The balance is fine-tuned until both the prosthetist and the amputee are satisfied (Figure 7.5).

### 7.2.6   Gait Training

A therapist gives the patient prosthetic training to assess the quality of the transtibial prosthesis (Figure 7.6). The human walk was divided into two phases according to the findings of a gait study: (1) the stance phase, when the foot is in contact with the ground; and (2) the swing phase, when the foot is not in contact with the ground (i.e., it is in the air).

**FIGURE 7.4**   Procedures for making a definitive prosthesis.

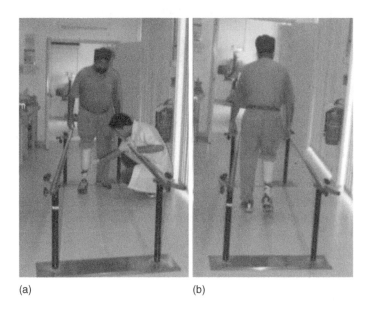

(a)                                          (b)

**FIGURE 7.5**   (a) Static alignment and (b) dynamic alignment.

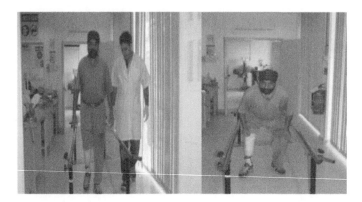

**FIGURE 7.6**   Gait training of the amputee.

## 7.3   THE INTERFACE PRESSURE IN TRANSTIBIAL SOCKET WITH DERMO AND SEAL-IN X5 LINERS

Background: The interface pressure between the residual limb and the prosthetic socket has a direct impact on the satisfaction and comfort of an amputee. Liners incorporate a soft cushion between the residual limb and the socket to provide a supportive interface. The Seal-In X5 and Dermo liner are two new interface systems and since they are still in their early stages, slightly are understood about their consequence on patient satisfaction. The goal was to evaluate the interface pressure with these two liners as well as their impact on patient satisfaction.

Methods: The research included nine unilateral transtibial amputees. Each amputee received two prostheses, one with the Seal-In X5 liner and one with the Dermo liner. During walking on level ground, interface pressure was assessed in the anterior, posterior, medial and lateral regions. Each participant completed a Prosthetic Evaluation Questionnaire (PEQ) to rate their satisfaction with the two liners.

Findings: The MPPs were 34% higher in the anterior, 24% higher in the posterior and 7% higher in the medial regions of the socket with the Seal-In X5 liner (P = 0.008, P = 0.046, P = 0.025) than with the dermo liner. At the lateral areas, there were no major variations in mean peak pressures between the two liners. Furthermore, a substantial difference in satisfaction and problems was discovered between the two liners (P <.05).

Interpretation: With the Dermo liner, there was less interface strain between the socket and the residual limb. According to the findings, the Dermo liner is more comfortable in the socket than the Seal-In X5 liner.

### 7.3.1   INTERFACE PRESSURE

Pressure measurements were taken in 12 different areas of the residual limb. Figure 7.7 depicts the stresses in the four main regions of the residual limb. The mean pressures for the proximal, middle sub-region areas were higher with the Seal-In X5 liner in both the anterior and posterior regions than with the Dermo liner (Figure 7.8).

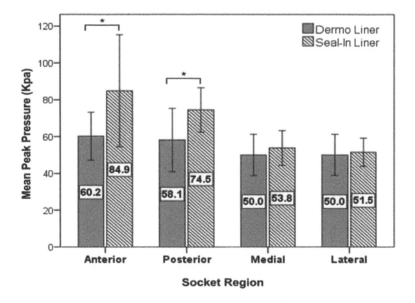

**FIGURE 7.7**   Mean peak pressure over the four main regions of the residual limb. Significant variations between the Dermo and Seal-In X5 liners are shown by the asterisks (*).

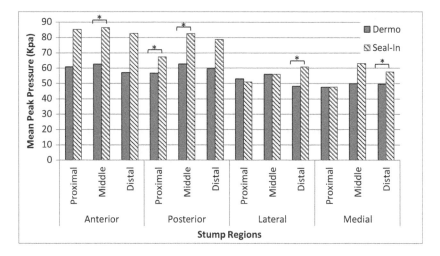

**FIGURE 7.8**   Mean peak pressure in all residual limb subsections. Significant variations between the Dermo and Seal-In X5 liners are shown by the asterisks (*).

**TABLE 7.1**

**Mean Peak Pressure (kPa) at the Anterior, Posterior, Medial and Lateral Sub-regions**

| | Anterior | | | Posterior | | |
|---|---|---|---|---|---|---|
| Liner Type | Proximal | Middle | Distal | Proximal | Middle | Distal |
| Dermo liner | 60.9(19.1) | 62.7(11.5) | 57.0(14.4) | 56.6(12.7) | 62.8(23.2) | 59.7(25.6) |
| Seal-In X5 Liner | 85.3(31.3) | 86.5(29.6) | 82.8(35.4) | 67.4(11.9) | 82.7(22.7) | 78.8(26.2) |
| P-value | 0.038* | 0.021* | 0.011* | 0.046* | 0.028* | 0.260 |
| Z | −2.07 | −2.31 | −2.54 | −1.99 | −2.19 | −1.125 |
| | Medial | | | Lateral | | |
| Liner Type | Proximal | Middle | Distal | Proximal | Middle | Distal |
| Dermo Liner | 47.6(13.9) | 49.9(12.8) | 49.5(19.0) | 53.0(26.3) | 56.1(14.5) | 48.2(9.4) |
| Seal-In X5 liner | 47.7(10.2) | 63.0(17.3) | 57.6(17.5) | 51.0(28.7) | 56.1(5.8) | 60.8(17.2) |
| P-value | 0.674 | 0.008* | 0.028* | 0.767 | 0.889 | 0.093 |
| Z | −0.42 | −2.66 | −2.19 | −0.29 | −0.14 | −1.68 |

* Significant differences between the Dermo and Seal-InX5 liner.

### 7.3.2 QUESTIONNAIRE

The Wilcoxon Signed Rank test showed substantially higher scores across five items for the Dermo liner and two items (including pistoning inside the socket and unexpected sounds) for the Seal-In X5 liner in the questionnaire feature aimed at evaluating system problems (Table 7.1).

### 7.3.3 DISCUSSION

The MPP at the posterior proximal area was 56.6 kPa for the Dermo liner and 67.4 kPa for the Seal-InX5 liner, respectively. There was a statistically significant difference between the two liners on the entire medial region of the residual limb in the current study, but no statistically significant differences between the two liners on the entire lateral region of the residual limb were reported.

Subjects were more pleased with the Dermo liner (Table 7.2), with the average difference across the nine questions on the satisfaction scale of the questionnaire being 8.67% higher for the Dermo liner and the mean difference was 4.69% higher for the Dermo liner than the Seal-In X5 liner. Both variations were statistically important.

The participants in this study find doffing and donning the Seal-In X5 liner much more difficult than for the Dermo liner. Because of difficulties donning and doffing the system and the extreme tightness of the socket, the participants in this sample, all of whom were over 50 years old, were not ready to embrace the

## TABLE 7.2
## Satisfaction and Problems with Dermo and the Seal-In X5 Liner

| | Dermo Liner Mean (SD) | Seal-In Liner Mean (SD) | P-value | Z | Effect Size |
|---|---|---|---|---|---|
| **Satisfaction** | | | | | |
| Fit of prosthesis | 78.1(5.6) ↑ | 73.3(5.6) | 0.011* | −2.46 | 0.58 |
| Ability to don and doff the prosthesis | 86.7(7.9) ↑ | 50.0(7.1) | 0.011* | −2.68 | 0.63 |
| Ability to sit with the prosthesis | 77.2(7.1) | 75.6(5.3) | 0.47 | NS† | — |
| Ability to walk with the prosthesis | 84.2(5.3) ↑ | 76.1(5.5) | 0.013* | −2.72 | 0.64 |
| Ability to walk on uneven terrain | 75.8(6.4) ↑ | 72.8(5.7) | 0.034* | −2.12 | 0.50 |
| Ability to walk up and down on stairs | 75.0(9.4) | 77.8(6.2) | 0.251 | NS | — |
| Suspension | 82.2(3.6) | 85.6(5.8) ↑ | 0.032* | −2.12 | 0.50 |
| Appearance of the prosthesis | 81.4(5.1) | 83.9(4.2) | 0.133 | NS | — |
| Overall satisfaction with the prosthesis | 84.7(5.7) ↑ | 70.6(4.6) | 0.015* | −2.09 | 0.49 |
| Overall Score | 80.6(5.1) ↑ | 73.9(4.0) | 0.004* | t = 9.02 | 0.91 |
| **Problems/Complaints** | | | | | |
| Sweating | 76.7(6.6) | 73.9(9.6) | 0.494 | NS | — |
| Wounds/ingrown hairs/ blisters | 87.8(7.9) ↑ | 82.2(7.9) | 0.041* | −2.06 | 0.49 |
| Skin Irritations | 84.4(8.8) ↑ | 77.2(9.7) | 0.041* | −2.03 | 0.48 |
| Pistoning within the socket | 78.9(6.0) | 86.7(5.6) ↑ | 0.013* | −2.14 | 0.50 |
| Rotation within the socket | 84.6(8.1) | 82.8(9.1) | 0.464 | NS | — |
| Swelling of the residual limb | 87.8(6.2) ↑ | 78.6(8.4) | 0.013* | −2.54 | 0.60 |
| Unpleasant smell of prosthesis or residual limb | 82.8(7.5) ↑ | 74.4(4.6) | 0.024* | −2.39 | 0.56 |
| Unwanted sounds | 77.8(3.6) | 83.9(4.9) ↑ | 0.015* | −2.42 | 0.57 |
| Pain in residual limb | 86.7(4.3) ↑ | 73.0(8.0) | 0.013* | −2.71 | 0.64 |
| Overall Score | 83.0(4.6) ↑ | 79.2 (5.9) | 0.012* | t = 3.20 | 0.57 |

* Significant differences between the Dermo and Seal-In liner.
† Non-significant.

Seal-In X5 liner. Except for suspension, the satisfaction score was higher for the Dermo liner with shuttle lock than for the Seal-In X5 liner. Furthermore, statistical analysis revealed that the Dermo liner had substantially less issues with the shuttle lock.

## 7.4 INTERFACE PRESSURE IN SOCKET DURING ASCENT AND DESCENT ON STAIRS

Background: Excessive pressure between the residual limb and the socket can impair prosthetic users' ability to walk up and down stairs. The study's goals were to compare the interface pressures of the Dermo (shuttle lock) and Seal-In (prosthetic valve) interface systems during stair ascent and descent, as well as to assess user satisfaction.

Methods: The research included ten amputees with unilateral amputations. During stair ascent and descent at a self-selected speed, interface pressure was measured using an F-socket transducer (9811E). Each participant completed a questionnaire about satisfaction and issues with the two interface systems.

Findings: During stair ascent, the Dermo interface system produced significantly lower mean peak pressure (kPa) than the Seal-In X5 interface system in the anterior, posterior, and medial regions (63.14 vs. 80.14, 63.14 vs. 90.44, 49.21 vs. 66.04, respectively) and incline (67.11 vs. 80.41, 64.12 vs. 88.24, 47.33vs. 65.11, respectively). In terms of satisfaction and problems faced, there was a significant statistical disparity between the two interface schemes ($P < 0.05$).

### 7.4.1 INTERFACE PRESSURE DURING ASCENT

The magnitude of the MPP was substantially greater in the posterior region ($P = 0.031$, $Z = -2.09$) with the Seal-In X5 (mean = 90.44 kPa, SD = 46.34) compared to the Dermo (mean = 63.13, SD = 9.21). Moreover, MPP was significantly higher in the anterior region ($P = 0.002$, $Z = -2.80$) with the Seal-In X5 (mean = 80.14 kPa, SD = 18.01) compared to the Dermo (mean = 63.14 kPa, SD = 13.40). Significant difference ($P = 0.031$, $Z = -2.09$) in Seal-In X5 interface also (mean = 66.04 kPa, SD = 30.22) compared to the Dermo (mean = 49.21 kPa, SD = 8.03) at the medial area. Yet, a major difference was found in the residual limb's medial distal sub-region.

### 7.4.2 INTERFACE PRESSURE DURING DESCENT

Regarding satisfaction, higher scores to the Dermo interface system compared to the Seal-In X5 interface system for three out of the four questions. However, the Seal-In X5 interface system obtained higher score for the suspension of the prosthesis with the residual limb during stair negotiation. Overall satisfaction was significantly higher for the Dermo interface system compared to the Seal-In X5 interface system (Table 7.3). The participants reported less pain with the Dermo interface system unlike the Seal-In X5 interface system (Table 7.4).

### 7.4.3 DISCUSSION

The current study found that the MPP was substantially higher at the anterior, posterior, and medial regions with the Seal-In X5 interface system compared to the Dermo interface system during both stair ascent and descent (24.72%, 35.56% and 29.20%,

**TABLE 7.3**

**Dermo and Seal-In X5 Interface Device Satisfaction during Stair Ascent and Descent**

| Stair Ascent | | | |
|---|---|---|---|
| Satisfaction Type/Interface Type | Mean | *P*-value | Z |
| Walking satisfaction during stairs ascent | 84.50 | 0.002* | −0.86 |
| Dermo | 72.90 | | |
| Seal-In X5 | | | |
| Suspension satisfaction during stairs ascent | 72.50 | 0.014* | −2.37 |
| Dermo | 82.13 | | |
| Seal-In X5 | | | |
| Balance satisfaction during stairs ascent | 78.00 | 1.006 | 0.00 |
| Dermo | 78.00 | | |
| Seal-In X5 | | | |
| Overall satisfaction during stairs ascent | 78.30 | 0.024* | −2.32 |
| Dermo | 72.50 | | |
| Seal-In X5 | | | |
| Stair Descent | | | |
| Satisfaction/Interface Type | Mean | *P*-value | Z |
| Walking satisfaction during stairs descent | 85.00 | 0.005* | −1.03 |
| Dermo | 70.50 | | |
| Seal-In X5 | | | |
| Suspension satisfaction during stairs descent | 75.20 | 0.002* | −2.69 |
| Dermo | 85.21 | | |
| Seal-In X5 | | | |
| Balance satisfaction during stairs descent | 75.20 | 0.313 | −1.00 |
| Dermo | 76.33 | | |
| Seal-In X5 | | | |
| Overall satisfaction during stairs descent | 84.20 | 0.014* | −2.53 |
| Dermo | 76.20 | | |
| Seal-In X5 | | | |

* Significant differences between the Dermo and Seal-In X5 interface system.

respectively). MPP was lower with the Dermo interface system in both the proximal and distal sub-regions as compared to the Seal-In X5 interface system.

During stair ascent, pressure was significantly higher at the proximal socket region, including the patellar tendon, according to this report. The magnitude of pressure was greater in our analysis at the posterior proximal region. During stair descent, participants in the current study faced more pressure at the anterior distal region with the Seal-In X5 interface system than with the Dermo interface system (Figure 7.9).

The amount of pressure produced by the two interface systems differed significantly. A relation maybe happens between low pistoning and higher MPP with the Seal-In X5. The amount of pistoning will decrease as the socket fit improves. Thus, tight fit of Seal-In X5 socket might be associated with lower pistoning. the close fit

**TABLE 7.4**

**In Terms of Problem Detection, a Comparison of the Dermo and Seal-In X5 Interface Systems during Stair Ascent and Descent Is Made**

| Stair Ascent | | | |
|---|---|---|---|
| Problem Type/Interface Type | Mean | P-value | Z |
| Pain during stair ascent | 87.00 | 0.005* | −2.67 |
| Dermo | 64.10 | | |
| Seal-In X5 | | | |
| Pistoning during stair ascent | 72.00 | 0.142 | −1.47 |
| Dermo | 76.50 | | |
| Seal-In X5 | | | |
| Rotation of the socket during stair ascent | 85.50 | 0.484 | −0.70 |
| Dermo | 86.50 | | |
| Seal-In X5 | | | |
| **Stair Descent** | | | |
| Problem Type/Interface Type | Mean | P-value | Z |
| Pain during stair descent | 78.00 | 0.011* | −2.55 |
| Dermo | 70.00 | | |
| Seal-In X5 | | | |
| Pistoning during stair descent | 74.50 | 0.171 | −1.36 |
| Dermo | 79.00 | | |
| Seal-In X5 | | | |
| Rotation of the socket during stair descent | 85.50 | 0.482 | −0.70 |
| Dermo | 86.50 | | |
| Seal-In X5 | | | |

* Significant differences between the Dermo and Seal-In X5 interface system.

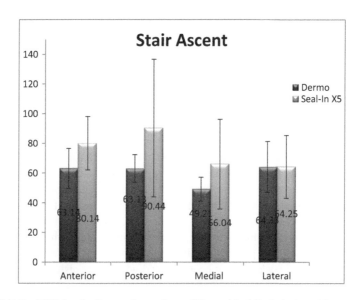

**FIGURE 7.9** MPP for the four major regions of the residual limb during stair ascent.

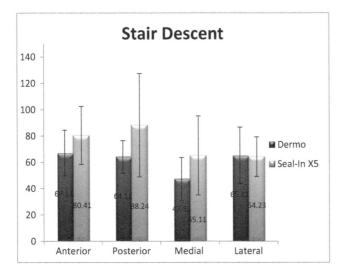

**FIGURE 7.10**  During stair descent, MPP for the four main regions of the residual limb.

has resulted in increased stress at the interface, which can be detrimental to the residual limb.

The transducers were very small and making positioning between the residual limb and interface device easier and protected more than 90% of the residual limb for a complete pressure map. This transducer offers a better sketch of residual limb pressure than single-spot transducers and can provide extra useful details for the clinical assessment of pressure-related complications.

The findings showed a substantial difference in the satisfaction and problems found by participants who used the two distinct prosthetic interface systems. When opposed to the Seal-In X5 interface, the participants had less issues with the Dermo interface system. Total satisfaction with the Dermo interface system was slightly higher (8.01%), and participants had less issues with the Dermo interface system (9.97%) (Figure 7.10).

## 7.5  THE EFFECT OF DERMO AND SEAL-IN X5 PROSTHETIC LINERS ON PRESSURE DISTRIBUTIONS

Background: Lower limb amputees have a difficult time navigating external obstacles such as ramps and stairs, and they have recorded high interface stress between the residual limb and socket/liner. Interface pressure between the residual limb and the socket/liner may influence prosthesis satisfaction. Yet, less awareness has been given to the interface pressure between the socket and the residual limb during ramp negotiation and its effect on amputee satisfaction.

Aim and Design: The objective was to compare the interface pressure delivered by two different liners (Seal-In X5 and Dermo) between the residual limb and socket, as well as their impacts on amputee satisfaction during ramp negotiation.

Methods: For each amputee, two prostheses were made. After four weeks of acclimation, the interface pressure between the socket and the residual limb was assessed when walking on the ramp and completed the PEQ for each liner.

Results: In ramp walking, MPP was lower with the Dermo liner compared to the Seal-In X5 liner. Dermo liner was preferred by the participants (83.50 vs. 71.50) and fewer issues (87.00 vs. 69.00) during ramp negotiation compared with the Seal-In X5 liner.

Conclusion: Because of the relative decrease in interface pressure, satisfaction and fewer problems, it is possible to assume that Dermo liner may be a good option for transtibial level amputation.

## 7.5.1 Discussion

Using the Seal-In X5 liner, both participants experienced increased pressure, with major variations observed in the anterior, posterior and lateral regions during the ramp ascent. The statistics during ramp descent revealed major variations at the anterior, posterior and medial ends. Furthermore, the participants were happier and had less issues with the Dermo line.

The questionnaire findings showed a preference for the Dermo over the Seal-In X5 liner. The Dermo liner made the participants' residual limbs stronger, and they were more comfortable than with the Seal-In X5 suspension system. The participants reported that when wearing the Dermo liner, their ability to ambulate was dramatically improved.

To achieve prosthesis fit, silicone liners are rolled over the residual limb. The Dermo liner made donning and doffing much simpler. Of the subjects, 90% registered difficulty in donning and doffing with the Seal-In X5 liner, and that doing so was extremely irritating. These difficulties may be due to the five seals found around

**TABLE 7.5**
**Dermo and Seal-In X5 Liners: Satisfaction and Issues**

| Description | Dermo (SD) | Seal-In X5 (SD) | P-Value |
|---|---|---|---|
| Fitting satisfaction | 83.10 (8.130) | 76.20 (7.81) | 0.005 |
| Donning/Doffing satisfaction | 86.00 (8.75) | 65.50 (7.61) | 0.003 |
| Ramp-down satisfaction | 83.50 (5.79) | 77.00 (5.37) | 0.001 |
| Ramp-up satisfaction | 79.00 (9.36) | 77.50 (10.60) | 0.391 |
| Overall satisfaction | 83.50 (8.54) | 71.50 (7.47) | 0.002 |
| Sounds problem | 72.50 (10.06) | 74.00 (8.43) | 0.341 |
| Pain problem | 84.00 (5.16) | 72.00 (6.32) | 0.004 |
| Sweating problem | 78.00 (7.52) | 68.80 (10.85) | 0.003 |
| Pistoning problem | 75.00 (5.50) | 75.00 (7.81) | 0.842 |
| Rotation problem | 80.00 (8.16) | 82.00 (6.32) | 0.261 |
| Smell problem | 77.50 (7.16) | 78.00 (7.52) | 0.591 |
| Overall problems | 87.00 (6.32) | 69.00 (5.67) | 0.003 |

Satisfaction: 100 indicated "completely satisfied" and 0 represented "unsatisfied".
Problem: 100 represented "not bothered at all" and 0 indicated "extremely bothered".

the liner, which create friction and do not slip easily unless lubricating spray is used. A few participants indicated more secure ramp negotiation with the Seal-In X5 liner due to the firm contact with the residual limb.

In short, the Dermo liner greatly reduced MPP during ramp descent and ascent (Table 7.5). As compared to the Seal-In X5 liner, both participants recorded being more involved and mobile with the Dermo liner; they could walk for longer periods of time with the Dermo liner.

## 7.6   COMPARATIVE STUDY BETWEEN DERMO, SEAL-IN X5 AND PELITE LINERS

Aim: The aim was to compare the effect of Dermo with shuttle lock, Seal-In X5 and pelite liners on amputee satisfaction and perceived problems.

Material and Methods: Thirty amputees (17 men and 13 women) volunteered to take part. For each participant, two prostheses were created. The participants completed the PEQ with the three liners.

Results: Dermo liners scored significantly higher in walking, walking on rough surfaces, stair walking, fitting, donning/doffing, sitting, suspension and overall satisfaction as compared to Seal-In X5 and Pelite liners. Overall satisfaction with Dermo liner was 34% higher than with Seal-In X5 liner and 28% higher than with pelite liner. Participants experienced less issues with the Dermo liner, and substantial variations in sweating, skin discomfort, annoyance and pain were observed between the three liners as compared to the Seal-In X5 and Pelite liners.

Conclusion: Dermo liner provided participants with a high degree of satisfaction and fewer problems. These findings indicate that the Dermo liner is a reasonable option for users and should assist clinicians and prosthetic practitioners in selecting requirements for prosthetic liners.

### 7.6.1   COMPARISON BETWEEN DERMO AND SEAL-IN X5 LINERS

In 7 of 10 questions, participants demonstrated differences between Dermo and Seal-In X5 liners. When comparing Dermo liners to Seal-In X5 liners, participants recorded 43.71% higher satisfaction during donning and doffing and 43.82% higher satisfaction during level walking with Dermo liners. As compared to Seal-In X5 liners, satisfaction with the Dermo liner was 50.34% higher when feeling with the prosthesis and 29.45% higher when walking on rough surfaces. Sum up, Dermo liner was 29.72% more pleased by participants than Seal-In X5 liner.

### 7.6.2   COMPARISON BETWEEN DERMO AND PELITE LINER

Participants were more pleased with the Dermo liner than the pelite liner, and there were major differences during prosthesis fit (87.00 vs. 74.00; $P = 0.002$, respectively), donning/doffing (92.00 vs. 87.00; $P = 0.051$, respectively), sitting with the prosthesis (90.00 vs. 83.00; $P = 0.033$, respectively), walking with prosthesis (89.00 vs. 78.00; $P = 0.004$, respectively), walking on rough surfaces (74.00 vs. 66.50;

$P = 0.007$, respectively), feel with the prosthesis (92.00 vs. 669.00; $P = 0.004$, respectively) and suspension with the prosthesis (88.50 vs. 74.50; $P = 0.011$, respectively). There are no variations in the appearance or weight of the prosthesis. Dermo liner provided 25.16% more overall satisfaction than pelite liner. During residuum skin discomfort, the Dermo liner scored higher than the pelite liner (90.00 vs. 75.00; $P = 0.001$, respectively), pain (90.00 vs. 70.00; $P = 0.005$, respectively) and dissatisfaction with the prosthesis (90.00 vs. 71.00; $P = 0.003$, respectively). Sweating was greatly reduced when using pelite liner versus Dermo liner (92.00 vs. 76.00; $P = 0.001$, respectively). The wound, sound and smell between the two liners were not statistically significant.

### 7.6.3 COMPARISON BETWEEN SEAL-IN X5 AND PELITE LINER

During donning/doffing, participants were significantly more satisfied with the pelite liner than with the Seal-In X5 liner (59.00 vs. 87.00; $P = 0.002$, respectively), walking (57.00 vs. 78.00; $P = 0.004$, respectively), walking on uneven surfaces (55.00 vs. 66.50; $P = 0.003$, respectively) and feel with the prosthesis (55.00 vs. 69.00; $P = 0.002$, respectively).

Suspension was noticeably improved with the Seal-In X5 liner. Concerning the problems encountered with the two liners, significantly less sweating (92.00 vs. 67.00; $P = 0.023$, respectively) reported with pelite liner and fewer pain (80.00 vs. 70.00; $P = 0.032$, respectively) detected with Seal-In X5 liner. There was no difference in swelling, scent or wound between the two liners.

### 7.6.4 DISCUSSION

Participants preferred the Dermo liner with shuttle lock over the pelite liner and Seal-In X5 liner. The participants reported has greater satisfaction while driving, walking on stairs and walking on rough ground while wearing the locking liner. Skin inflammation, ulcers and abrasions are common problems associated with the use of prostheses and occur in amputees (Table 7.6).

These skin issues cause irritation and pain, and in some cases, amputees avoid using their prosthesis entirely. Still, the Dermo and Seal-In X5 liners caused more sweating than the pelite liners. With the Dermo liner, participants feel more satisfied and experience less discomfort, allowing them to walk more comfortably.

The fit of the socket and the suspension system of the prosthesis have a significant effect on the participant's comfort, satisfaction and mobility. Silicon liners are rolled over the residuum and tightly bound to the residuum's skin, forming a bond between the residuum and the liner. Seal-In X5 and Dermo liner suspension were preferred by the participants.

The ease of donning and doffing the prosthesis has a significant impact on prosthetic users. In the current research, the Dermo liner demonstrated significantly easier donning and doffing as compared to other liners. In this study, all participants indicated considerable difficulties donning and doffing the Seal-In X5 liner, which could be attributed to the five seals around the liner.

**TABLE 7.6**

**Dermo vs. Pelite Liner Comparison**

| | Satisfaction | | |
|---|---|---|---|
| **Variable** | **Dermo** | **Pelite** | **P-value** |
| Fit of prosthesis | 87.00 (2.53) | 74.00 (4.10) | 0.002 |
| Donning/doffing | 92.00 (2.63) | 87.00 (8.12) | 0.051 |
| Sitting with prosthesis | 90.00 (6.70) | 83.00 (6.32) | 0.033 |
| Walking with prosthesis | 89.00 (5.17) | 78.00 (5.37) | 0.004 |
| Walking on Uneven surface | 74.00 (5.70) | 66.50 (4.74) | 0.007 |
| Walking on stairs | 67.00 (5.35) | 62.50 (5.40) | 0.003 |
| Appearance of prosthesis | 87.00 (2.73) | 85.00 (4.71) | 0.336 |
| Feel with prosthesis | 92.00 (2.73) | 69.00 (7.10) | 0.004 |
| Weight of prosthesis | 86.00 (5.47) | 86.00 (5.16) | 1.002 |
| Suspension | 88.50 (7.07) | 74.50 (11.41) | 0.011 |
| Overall satisfaction | 85.00 (2.5) | 66.00 (11.25) | 0.007 |
| | Problems | | |
| **Variables** | **Dermo** | **Pelite** | **P-value** |
| Sweating | 76.00 (5.1) | 92.00 (5.37) | 0.001 |
| Sound | 80.00 (6.66) | 80.00 (6.66) | 1.002 |
| Skin irritation | 90.00 (0.00) | 75.00 (7.45) | 0.001 |
| Smell | 78.00 (8.11) | 75.00 (8.81) | 0.642 |
| Wound | 92.00 (8.10) | 85.00 (9.42) | 0.082 |
| Pain | 90.00 (2.10) | 70.00 (5.77) | 0.005 |
| Frustration | 90.00 (3.33) | 71.00 (6.14) | 0.003 |
| swelling | 88.00 (2.60) | 85.00 (4.71) | 0.130 |

Satisfaction: 100 represented "completely satisfied" and 0 indicated "not satisfied at all".
Problem: 100 represented "not bothered at all" and 0 indicated "extremely bothered".

In conclusion, all the participants were pleased with the Dermo liner and demonstrated excellent performance when walking on level, stair and rough surfaces. The findings also show that the Dermo liner caused less difficulties and frustrations for the participants (Table 7.7).

**TABLE 7.7**

**Comparison between Seal-In X5 and Pelite Liner**

| | Satisfaction | | |
|---|---|---|---|
| **Variable** | **Seal-In X5** | **Pelite** | **P-value** |
| Fit of prosthesis | 78.00 (2.73) | 74.00 (4.10) | 0.021 |
| Donning/doffing | 59.00 (12.41) | 87.00 (8.12) | 0.002 |
| Sitting with prosthesis | 87.00 (4.27) | 83.00 (6.32) | 0.141 |
| Walking with prosthesis | 57.00 (12.70) | 78.00 (5.37) | 0.004 |
| Walking on Uneven surface | 55.00 (8.35) | 66.50 (4.74) | 0.003 |
| Walking on stairs | 62.00 (7.17) | 62.50 (5.40) | 0.933 |
| Appearance of prosthesis | 87.00 (2.73) | 85.00 (4.71) | 0.334 |

*(Continued)*

### TABLE 7.7 (*Continued*)

| | Satisfaction | | |
|---|---|---|---|
| Variable | Seal-In X5 | Pelite | *P*-value |
| Feel with prosthesis | 55.00 (11.24) | 69.00 (7.10) | 0.002 |
| Weight of prosthesis | 86.00 (5.47) | 86.00 (5.16) | 1.000 |
| Suspension | 91.00 (7.37) | 74.50 (11.41) | 0.004 |
| Overall satisfaction | 63.00 (7.91) | 66.00 (11.25) | 0.531 |
| | Problem | | |
| Variables | Seal-In X5 | Pelite | *P*-value |
| Sweating | 67.00 (2.58) | 92.00 (5.37) | 0.023 |
| Sound | 82.00 (8.10) | 80.00 (6.66) | 0.515 |
| Skin irritation | 83.00 (5.41) | 75.00 (7.45) | 0.022 |
| Smell | 78.00 (7.91) | 75.00 (8.81) | 0.341 |
| Wound | 90.00 (9.42) | 85.00 (9.42) | 0.202 |
| Pain | 80.00 (3.33) | 70.00 (5.77) | 0.032 |
| Frustration | 71.00 (7.10) | 71.00 (6.14) | 0.872 |
| swelling | 88.00 (4.21) | 85.00 (4.71) | 0.113 |

Satisfaction: 100 represented "completely satisfied" and 0 indicated "not satisfied at all".
Problem: 100 represented "not bothered at all" and 0 indicated "extremely bothered".

## 7.7 SATISFACTION OF INDIVIDUALS WITH AMPUTATION AND PROBLEMS WITH THEIR PROSTHETIC DEVICES

Objective: The aim of this study was to investigate the effects of three different suspension systems, namely the polyethylene foam liner, the silicon liner with shuttle lock, and the Seal-In liner, on the satisfaction and perceived problems with participants' prostheses.

Participants: A total of 243 people with unilateral amputations were fitted with prostheses that included a polyethylene foam liner, a silicon liner with shuttle lock, and a Seal-In liner.

Main Outcome Measure: Descriptive studies were conducted on the study participants' demographic details, happiness, and prosthesis-related problems.

Results: The findings revealed substantial differences in satisfaction and perceived issues with the prosthetic device among the three classes. Individual item analyses showed that the study participants were more pleased with the Seal-In liner and had less issues with it. Users of the silicon liner with shuttle lock and the Seal-In liner recorded significantly shorter maintenance times as compared to the polyethylene foam liner. Users of the silicon liner with shuttle lock sweated more, while those who used the Seal-In liner had more difficulty donning and doffing the unit.

Conclusion: The survey findings indicate that the Seal-In liner improves prosthetic suspension as compared to the polyethylene foam liner and silicon liner with shuttle lock. Though, further prospective studies are needed to determine which method offers the most comfort and the fewest problems for participants.

## 7.7.1  DISCUSSION

The results confirmed our hypothesis that participants will be more comfortable with the Seal-In liner than with the other two schemes (Table 7.8). Except for the "sweat complaint", major variations in perceived problems were discovered between different suspension systems. PTAs with the locking liner (55 score) indicated sweating more frequently than those with the polyethylene foam and Seal-In liners.

### TABLE 7.8
### In Terms of Complaints/Problems, a Comparison of Three Different Suspension Systems Is Made

| Problem/Liner Type | Mean*P-value | Ranking† |
|---|---|---|
| **Sweat complaint** | .074 | |
| Silicone liner with shuttle lock | 55.00 | 3 |
| Polyethylene foam liner | 60.16 | 2 |
| Seal-In liner | 64.78 | 1 |
| **Wound complaint** | .002 | |
| Silicone liner with shuttle lock | 81.85 | 2 |
| Polyethylene foam liner | 75.04 | 3 |
| Seal-In liner | 95.17 | 1 |
| **Irritation complaint** | .005 | |
| Silicone liner with shuttle lock | 81.28 | 2 |
| Polyethylene foam liner | 75.10 | 3 |
| Seal-In liner | 94.66 | 1 |
| **Pistoning within the socket** | .006 | |
| Silicone liner with shuttle lock | 84.18 | 2 |
| Polyethylene foam | 63.95 | 3 |
| Seal-in liner | 96.47 | 1 |
| **Rotation within the socket** | .002 | |
| Silicone liner with shuttle lock | 80.18 | 3 |
| Polyethylene foam liner | 81.65 | 2 |
| Seal-In liner | 99.57 | 1 |
| **Inflation complaint** | .021 | |
| Silicone liner with shuttle lock | 86.75 | 2 |
| Polyethylene foam liner | 89.64 | 3 |
| Seal-In liner | 94.91 | 1 |
| **Smell complaint** | .004 | |
| Silicone liner with shuttle lock | 72.49 | 2 |
| Polyethylene foam liner | 63.94 | 3 |
| Seal-In liner | 94.91 | 1 |
| **Sound complaint** | .003 | |
| Silicone liner with shuttle lock | 70.21 | 3 |
| Polyethylene foam liner | 80.28 | 2 |
| Seal-In liner | 96.81 | 1 |
| **Pain complaint** | .004 | |
| Silicone liner with shuttle lock | 80.62 | 3 |
| Polyethylene foam liner | 81.18 | 2 |
| Seal-In liner | 92.67 | 1 |

* Greater mean indicates more satisfaction and use.
† Satisfaction increases the ranking from 3 to 1.

Furthermore, we found substantial differences between suspension liners in terms of participant usage and satisfaction. The whole satisfaction score was greater with the Seal-In liner (83.10 score) when contrasted with the locking liner (75.94%) and the polyethylene foam liner (63.14 score). The silicone liner with shuttle lock and Seal-In liner was favoured by the participants over the polyethylene liner.

Donning and doffing was better for participants who used the polyethylene and locking liners than for those who used the Seal-In liner, according to the findings. According to the results, the polyethylene liner was the most robust of the three suspension systems.

This research did not focus solely on happiness and perceived issues. Likewise, we discovered that for the Seal-In liners, all the satisfaction parameters were higher than for the locking system and the polyethylene foam liner. Moreover, statistical analysis showed that the Seal-In liner caused less issues for the participants than the other two liners.

## 7.8 CONCLUSIONS

The results of interface pressure analyses revealed that wearing the Dermo liner when level walking results in less pressure inside the socket. Furthermore, the Dermo liner caused fewer problems and concerns among the subjects. As a result, we can assume that the Dermo liner offers a more secure socket–residual limb interface during level walking than the Seal-In X5 liner. Despite this, the Seal-In X5 liner provides improved suspension while level walking.

During their everyday activities, prosthesis users encounter a variety of routes, including level floors, ramps, stairs and other uneven surfaces. Because of the loss of the foot and ankle system, lower-limb amputees have a difficult time navigating environmental obstacles such as slopes and stairs and have reported high interface strain.

With the Seal-In X5 interface systems, this study found that there is a lot of strain between the residual limb and the socket. The Dermo interface system applied minimal pressure, and the participants felt more at ease when negotiating the stairwell. During stair negotiation, the participants felt more secure and at ease using the Dermo interface framework. The results of this study revealed that a high magnitude of pressure was registered with the Seal-In X5 liner during ramp ascent and descent.

Participants were more comfortable, had less issues and favoured the Dermo liner over the other liners when comparing Dermo, pelite and Seal-In X5 liners. For transtibial prosthetic users, dermo liner might be the best option.

## REFERENCES

Bui, K. M., Raugi, G. J., Nguyen, V. Q., & Reiber, G. E. (2009). Skin problems in individuals with lower-limb loss: Literature review and proposed classification system. *Journal of Rehabilitation Research & Development, 46*(9).

Chadderton, H. (1978). Prostheses, pain and sequelae of amputation, as seen by the amputee. *Prosthetics and Orthotics International, 2*(1), 12–14.

Hoxie, L. O. (1995). Outcomes measurement: a primer for orthotic and prosthetic care. *JPO: Journal of Prosthetics and Orthotics, 7*(4), 132–136.

Zhang, M., & Roberts, C. (2000). Comparison of computational analysis with clinical measurement of stresses on below-knee residual limb in a prosthetic socket. *Medical Engineering & Physics, 22*(9), 607–612.

# 8 Study of Pressure Feet between 3D Scanning Method and Conventional Casting

*N A Abu Osman*

University of Malaya, Kuala Lumpur, Malaysia

*W Mehmood*

St. Gabriel's Orthotics Crabtree House, Limerick, Ireland

*N A Abd Razak*

University of Malaya, Kuala Lumpur, Malaysia

## CONTENTS

DOI: 10.1201/9781003196730-8

## 8.1   INTRODUCTION: BACKGROUND

The different causes of lower limb loss may be due to injury caused by trauma, diabetes, cancer, congenital limb deficiency, or peripheral vascular disease (Michael and Bowker, 2004). Statistics shows that the number of diabetics is increasing worldwide with the passage of time, which in turn is showing dramatic change in the increasing percentage of lower limb amputation. Statistics shows that the main causes of amputation in the US is peripheral vascular disease and diabetes, comprising 54%, while trauma is about 45%, and cancer less than 2% of the total disabled population (Ziegler-Graham et al., 2008) Diabetes patients have the risk of experiencing lower limb amputation 10–30 times higher than the normal population who experience amputation (Vamos et al., 2010). Statistics also shows that transtibial amputation is the most common one in lower limb amputation comprising 47% of all lower limb amputation followed by transfemoral amputation which consists of 31% (Organization, 2004). After amputation, prosthetics are essential for individuals to regain the functional activity of walking (Wolf et al., 2009). The most important part of a prosthesis is the socket connecting the assembly of the prosthesis to the residual limb. The socket must fit and be comfortable to provide optimum gait without discomfort (Jia et al., 2004). The poor fitting of the socket will not only cause complications to the residual limb but also will affect the optimum gait of the individual (Robert Gailey, 2008).

The shape of residual limbs plays a vital role in fabricating sockets design and fitting the socket. There are three fundamental shapes of transtibial residual limb after amputation which are conical, cylindrical and bulbous. These differences may be due to the amputation technique performed by surgeons or residual limb management post amputation which causes differences in the condition of soft tissues, muscles and bone at the end of the residual limb.

There are many types of transtibial sockets fo0r amputees, but the main two types are Patellar Tendon Bearing (PTB) and Total Surface Bearing (TSB) sockets, as shown in Figure 8.1. In 1957 the first design in transtibial prosthesis became popular, which was the PTB design (Radcliffe, 1961). PTB sockets are triangular to avoid rotation of the sockets around the residual limb. The weight is taken mostly on the patellar tendon bar and posterior surface along with anterior tibial flare and lateral tibial flare. The tibial crest, anterior tibial prominence, tibial tuberosity, fibular head and end of residual limb is the pressure sensitive areas so need to be relieved. A PTB socket is usually prescribed with pelite liner and gel liners.

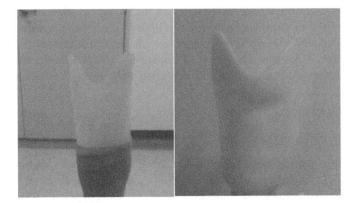

**FIGURE 8.1** PTB socket (L) and TSB socket (R).

TSB socket is the type of socket in which weight is equally distributed through the residual limb. In TSB socket the shape of the residual limb is followed where the reduction is done equally from all surfaces. The advantage of TSB socket is that there is no pressure on specific points on the residual limb unlike PTB socket which can be seen in Figure 8.2. PTB socket is usually prescribed with silicon liners with different systems for suspensions.

Manual casting using POP bandage, scanning using CAD CAM, sand casting, ICECAST system and many more. In manual casting, the POP bandage is wrapped around the residual limb, and shaped manually using hand of the residual limb. The negative cast is obtained and is filled using POP powder to get the positive cast which is the replica of the residual limb. The model is then modified using a SURFOM blade following the measurements and shape of the residual limb. The model is then laminated or draped to get the final socket. In scanning method, a hand help scanner is used to get the impression of the residual limb either by LASER scanning or image is taken. The scan is then opened in a software where necessary changes are done within that software to match the shape and measurements of the residual limb. the model is then printed, and socket is produced. In sand casting the residual limb is placed inside

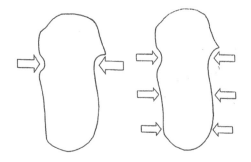

**FIGURE 8.2** Pressure distribution in a PTB socket (L) and TSB socket (R).

a bag full of sand which is connected to a suction. The air is sucked through the suction and the sand get the shape of the residual limb which is then filled with sand with a vacuum suction. The model is modified using hammer and sand with the suction on. After the desired shape is achieved POP bandage is used to take the cast from the model and filled with POP powder to be laminated or moulded for achieving a socket.

## 8.2  LITERATURE REVIEW

### 8.2.1  Lower Limb Amputation

Injury or disease occurs in the lower limb like road traffic accidents, vascular disease, diabetics, poor circulation and cancer can cause amputation in the lower limb. Apart from these acquired causes of amputation, some are also congenital. (Courvoisier et al., 2009). Vascular disease specifically diabetics comprised of 80–90% of total amputation occurs in western countries (Dillingham et al., 2002). Diabetes is a major issue in Malaysia, for those undergoing transtibial amputation. It is predicted by the International Diabetes Federation that the diabetic population will increase from 240 million in 2017 to 380 million in 2025, which will burden the lower and middle-income countries by 80%. The approximation shows that 60% of the number will come from Asia due to its populous contribution (Chan et al., 2009). According to an estimate given by the World Health Organization (WHO), the percentage of diabetics in Malaysia will increase by 10.8% until 2030, which is at very high risk. This increase may be due to the higher rate of obesity there (Mohamud et al., 2011).

### 8.2.2  Transtibial Prosthesis

Transtibial amputees after going through amputation needs to have a prosthesis which can replace the missing limb. Transtibial residual limbs are one of the complicated limbs because of the bony structure of the lower limb, so one needs to be precise when giving pressure and relief to required areas (Radcliffe, 1961). There are two bones, the tibia and fibula, present in the residual limb, which makes it more difficult in terms of creating a well-contoured socket (Jia et al., 2004). Some of the differences in pressure or relief may not give optimum fitting, and in fact offer discomfort or pain in the residual limb (Johansson and Öberg, 1998). The two basic types of transtibial prostheses are exoskeletal and endoskeletal prostheses. The exoskeletal prosthesis transfers the weight of the body through the exterior walls to the ground along its shape. Exoskeletal prostheses are protective and durable in humid conditions or rainy areas, and resist water. They are made of either plastic or wood and termed as "conventional prostheses".

The endoskeletal prosthesis can go with a variety of components including advanced technologies for different activity levels of the amputees. Endoskeletal provide a variety of components selection for amputees with varied activity level. Geriatric patients need stability, while young and athletic amputees need to do high impact activities, and it is well addressed by the endoskeletal system in selection of components.

Different parts of the transtibial prosthesis are a socket, soft in-liner, shank and foot. The socket is the part which is in contact with the body through the liner. This function makes socket the most important part of the prosthesis, that is why it needs to follow the shape of the residual limb in fabrication. The socket should have a balance of forces over pressure sensitive and pressure tolerant areas (Wolf et al., 2009).

The socket is the part which is in contact with the residual limb and connected to the pylon and all the way until foot. It is important to get a comfortable socket in a prosthesis for successful rehabilitation of the amputee (Goh et al., 2004). The forces in the socket are distributed equally according to the shape of the residual limb. Pressure is applied on the pressure tolerant areas while the sensitive areas need to be relieved. If the socket is not well-fitted then there will be discomfort and the residual limb will experience skin problems (Sanders et al., 2000). The symmetry is important in the gait as the prosthetic side must follow the gait parameters of the sound side or closer to it (Chow et al., 2006).

The PTB socket design follows a triangular shape with reliefs and pressure areas on the residual limb. The anterior wall is the most important wall which covers the patellar tendon which is the main weight bearing area in PTB socket design (Yiğiter et al., 2002). PTB socket design usually includes pelite liner. The patellar tendon bar is curved on the sides to make a smiley shape below the patella for weight bearing. The tibial crest is also covered in the anterior wall where relief is given on the anterior region. Posterior wall is an essential wall of the transtibial socket where a relief channel is provided for hamstrings tendon along with pressure on the gastrocnemius muscles and popliteal region for weight bearing (Goh et al., 2004). The pressure tolerant areas are patellar tendon, medial flare of tibia, lateral flare, and posterior region of the socket while the relived areas are fibular head, end of fibula, hamstrings tendon and end of tibia (Figure 8.3).

In TSB socket design the load is equally distributed all over the socket unlike PTB. The equal weight distribution all over the residual limb will help in good fit, more comfort, and good blood circulation (Staats and Lundt, 1987). There is no extra weight bearing on the patellar tendon and flares of the residual limb in TSB socket, but overall reduction is done from the model, and it follows anatomical shape. It was found by (Yiğiter et al., 2002) that TSB has more advantages then the PTB socket design.

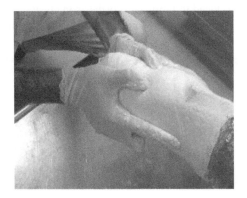

**FIGURE 8.3** Casting a transtibial residual limb.

### 8.2.3 Types of Fabrication Technique

The fabrication techniques which include manual casting with POP bandage, scanning and sand casting with suction (Öberg et al., 1993). Transtibial amputees with an optimum fitting of the prosthetic sockets function better than those having fitting problems of the prosthetic sockets (Bolt et al., 2010). Transtibial prostheses are fabricated using numerous methods which have been established to achieve the best possible socket. Socket's fabrications follow the basic principles with some modification and changes. The first socket fabricated was a PTB socket by (Radcliffe, 1961). There were eight factors listed as a guideline for properly fabricating the socket to provide details about the procedure and application on the socket.

### 8.2.4 Manual Casting Method

The manual casting using POP bandages as shown in Figure 8.4. It is important for a prosthetist to have the basic knowledge and skills to fabricate a well fitted socket. Residual limb of every amputee differs from another, so the socket also follows the same pattern. Prosthetist should have basic knowledge about pressure sensitive and pressure tolerant areas on the residual limb, so the prosthetist can follow them in the procedure for making a well fitted socket without any discomfort.

### 8.2.5 CAD CAM Socket

One of the techniques resulting was computer-aided designs and computer-aided manufacturing (CAD CAM). The CAD CAM technology in prosthetic fabrication involves creating a digital image of the residual limb which is stored in the computer and modified in the computer using a software. Once the adjustments are done the final image is sent to the carving machine where a three-dimensional mould is taken, which is a replica of the residual limb (Childress, 2002). CAD CAM was first introduced in 1965 and was refined later by many researchers to get different types of CAD CAM.

### 8.2.6 Sand Casting

Sand casting was introduced to fabricate prosthesis in cheaper price and its viability to the remote areas. Wu et al. (2003) conducted research to develop a fabrication

**FIGURE 8.4**   Casting a transtibial residual limb (L) and Modification of transtibial model (R).

system based on a dilatancy which is usually done in wheelchair fabrication. This system actually replaces the POP bandage and powder because both negative and positive casts are made of sand (Steen Jensen et al., 2005). Sand casting method involves a sack full of sand put on the residual limb, and negative suction is attached to end of the sack which is then opened to remove the air, and then filled, and the suction is opened again to remove the air and get the positive model.

### 8.2.7 ICECAST ANATOMY

The ICECAST anatomy is a new method known as modular socket system (MSS) for fabricating transtibial sockets, developed by OSSUR as shown in Figure 8.5. There is a casting bladder, shuttle lock components and silicon sleeve in the kit of ICECAST. It is believed that this is the fastest way to fabricate transtibial socket on the patient residual limb. The sleeves and the fibreglass are pulled over the residual limb. The resin and hardener are injected through a hole on the distal end of the residual limb. Once it is injected, the bladder is rolled over onto the residual limb and pressure is applied using the pressure gauge. This is one of the most expensive socket fabrications due to the cost of the components and materials.

### 8.2.8 3D SCANNING TECHNIQUE

The socket obtained using BioSculptor, scanning is done to take the impression of the residual limb by bioscanner as shown in Figure 8.6. There is a transmitter attached to the bioscanner which is fixed at one place near to the part which needs to be scanned. The LASER light strikes on the residual limb to take different sweeps which can be seen on the computer screen. The sweeps are taken all over the residual limb where we can get a full scan of the residual limb on the computer. The model can be viewed in 3D, so all the three planes can be viewed. Different bony marks are marked with the help of a stylus pen attached to the computer. After the 3D scan is completed then the file is saved, and the 3D model can be modified in BioShape software which can help us get the desired adjustments needed in the model and to alter the model if needed.

**FIGURE 8.5** Resin and hardener injected into the sleeves on residual limb (L), donning of casting bladder on residual limb (M) and casting bladder pumped on the residual limb (R).

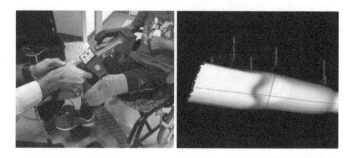

**FIGURE 8.6**  Scanning the residual limb using bioscanner (L) and Model modified using BioShape software (R).

## 8.3  METHODOLOGY

### 8.3.1  INTRODUCTION

Five patients were recruited for this study. For each patient, two sockets were fabricated using BioSculptor scanning method and conventional casting technique using POP bandages. The circumferential profile of the residual limb is considered as the control measure, compared with the circumferences of the positive model of the conventional and the 3D model of the BioSculptor model. Measurements are recorded twice, one prior to modification the other after.

### 8.3.2  PARTICIPANTS

As a sample of convenience five subjects were recruited from University Malaya Medical Centre (UMMC) after getting the ethical approval from the ethical committee of UMMC. The inclusion criteria for this study were unilateral transtibial amputees, mature and with well-formed residual limbs, aged from 20 to 50 years old, and of good mental condition. The patients selected for this experiment were those already using a transtibial prosthesis, and so aware of the procedures.

### 8.3.3  ETHICAL APPROVAL

This research is conducted with the approval of permission by National Medical Research Register Secretariat 37912 and under the guidance of Certified Prosthetist and Orthotist (CPO) ISPO CAT 1 (Table 8.1, Figure 8.7).

### 8.3.4  EXPERIMENTAL PROCEDURE: CONVENTIONAL CASTING TECHNIQUE

For the conventional technique, different landmarks are marked using on the patient residual limb which includes patella, patellar tendon, anterior tibial prominence, tibial tuberosity, fibular head, tibial crest, lateral tibial border, medical tibial border, hamstring tendons, fibular end and tibial end. The required measurement is taken, and the residual limb is positioned at 15-degree flexion for casting so that the bony prominences are easily palpated. All the required measurements are taken which

**TABLE 8.1**
**Showing the Demographic Data of All Patients**

| Sr. No | Age | Cause Amputation | Year of Amputation | Length of Residual Limb | Size of Residual Limb | Shape of Residual Limb | Soft Tissue Consistency | Side of Amputation | Activity Level |
|---|---|---|---|---|---|---|---|---|---|
| 1 | 52 | Diabetics | 2016 | 10 | Medium | Cylindrical | Firm | L | K2 |
| 2 | 59 | Diabetics | 2013 | 13 | Long | Cylindrical | Average | L | K2 |
| 3 | 28 | Trauma | 2011 | 17 | Long | Cylindrical | Average | L | K2 |
| 4 | 42 | Trauma | 2015 | 7 | Short | Cylindrical | Soft | L | K2 |
| 5 | 24 | Trauma | 2015 | 10 | Medium | Cylindrical | Firm | L | K4 |

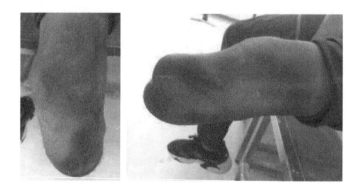

**FIGURE 8.7**   Anterior and lateral view of the residual limb.

includes circumferential measurements, length of the residual limb, medio-lateral diameter and anterio-posterior diameter. POP bandage is wrapped around the residual limb to take its impression to take a negative cast on the residual limb. pressure is applied on the patella tendon, popliteal fossa and the tibial flares until the cast get hard. Once the cast hardens it is removed from the residual limb to check for the shape which is called the negative cast (Figure 8.8).

The negative cast is filled using POP powder for producing a positive cast and a mandrill is inserted inside for holding later in the process. The plaster is removed once the POP powder gets harder and is kept in the bench vice.

The positive cast is modified using surform. The model is modified following the residual limb shape and the bony landmarks. The bony landmarks work as a guideline for removing or adding POP from the model. Pressure is applied on the pressure tolerant areas using surform blade by removing POP while relief is given to the pressure sensitive areas by adding POP on those areas. The pressure and relief on the model give the exact shape of the socket to provide good fit and comfort to the user (Figure 8.9).

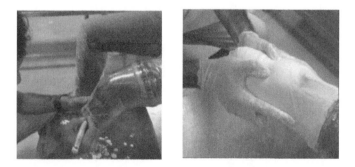

**FIGURE 8.8**   Landmarks marking and POP bandage casting.

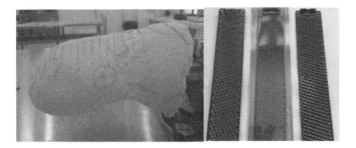

**FIGURE 8.9** Positive model and surform blades.

### 8.3.5 BioSculptor Technology

BioSculptor technology is a 3D CAD CAM technology for fabricating transtibial sockets (BioSculptor Innovative solutions, 1997c). It starts with scanning a residual limb using a 3D handheld bioscanner (BioSculptor Innovative solutions, 1997b) for taking the impression of the residuum using infrared light through sweeps on different parts of the residuum using digital imaging. There is a transmitter attached to the bioscanner which is fixed at one place near to the part which needs to be scanned (Figure 8.10).

After the 3D scan is completed, the file is saved. The model is then sent to a modification software known as BioShape which helps in doing the necessary modifications following the measurements recorded from the amputee. There are several different basic steps followed on BioShape initially, starting with matching measurements. All the necessary modifications are done to closely match the initial measurements and shape of the residual limb. The next step is to follow the shape of the residual limb and pressure is applied in pressure-tolerant areas, while relief is given on sensitive areas. This helps in comparing them to the measurements recorded earlier on the residual limb using conventional methods. The final step in BioShape is drawing the trim lines on the model, and the Biomill follows those guidelines for printing (Figure 8.11).

**FIGURE 8.10** Scanning the residual limb using bioscanner.

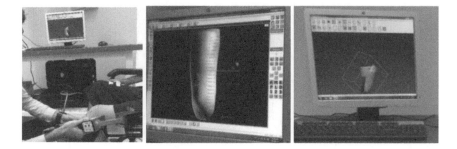

**FIGURE 8.11**   Getting 3D model after scanning.

The file is then saved as a cut file and is sent to the Biomill for printing. The positive model is obtained. The final model is then printed using Biomill which carves the model using a milling machine (Figures 8.12, 8.13, 8.14).

### 8.3.6   DATA ANALYSIS

The difference of the circumferential profile between BioSculptor (BS) and conventional casting of POP method is explained in this section where the control values are the residual limb circumference. The circumference is taken at each level from the supracondylar area at 1cm on the residual limb. The values are calculated on the same locations of the model using BioSculptor through scanning as well as the positive model obtained using conventional casting technique. The results are obtained in two parts which are without modification and after modification for both types of technique. The values are then analyzed using IBM SPSS Statistics 20 to find out the standard deviation and the significant values keeping confidence interval at 95%. There were different tests performed keeping objectives in mind.

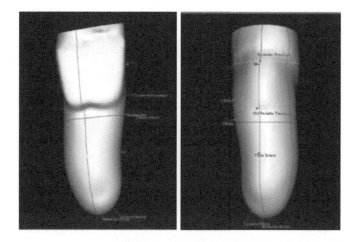

**FIGURE 8.12**   Modifications using BioShape manipulation software modification.

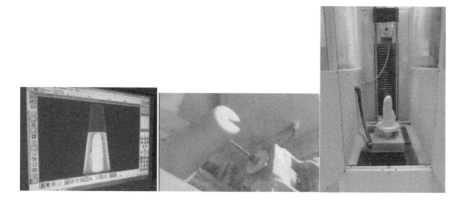

**FIGURE 8.13** Milling the 3D model using Biomill.

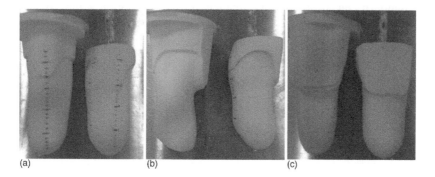

(a)                    (b)                   (c)

**FIGURE 8.14** (a) Anterior, (b) medial and (c) posterior view of Biomill printed and POP positive casts.

Paired t-test was applied to correspond the first objective to compare the circumferential profile of transtibial sockets fabrication using a conventional manual casting technique with a BioSculptor scanning technique. Pearson correlation was performed between the values to check the coalition profile difference at each layer of 1cm where pressure was applied at the residual limb using both prepared techniques.

## 8.4 RESULTS AND DISCUSSION

### 8.4.1 ANALYSIS BEFORE MODIFICATION

The values of the BS and the POP are plotted to check the deviation from the residual limb values. These values are recorded without any modifications in the models to see how much the values are deviated from the control values. The circumference is plotted on the y-axis while the length is plotted on x-axis. There are three lines which represents the circumferential values for residual limb, BioSculptor and POP. The residual limb values represented in blue colour are the control values of the residual limb of the patients, the red colour representing BS and green representing POP values before modification.

Statistical analysis was performed using SPSS to check the standard deviations and significant values of the BS and POP models comparing with the residual limb values as shown in Table 8.2.

Applying one sample statistic for a single pair test to find the significance, mean, standard deviation and confidence interval, it was found that the values are significant. The data is distributed normally, and this was checked using Shapiro–Wilk test for normality. Standard deviation among values differs from patient to patient as represented in the table. Table 8.3 indicates the Paired t-test was applied to further investigate the details of the accuracy in the circumferential profile of both BioSculptor and conventional models with reference to residual limb to correspond the first

### TABLE 8.2
### The Standard Deviation and Significance of Circumferential Values of the Residual Limb vs. BS and POP before Modifications Using One Sample t-test

| Before Modification | Residual Limb | | | BioSculptor Model | | | POP Model | | |
|---|---|---|---|---|---|---|---|---|---|
| | Mean | SD | P-Value | Mean | SD | P-Value | Mean | SD | P-Value |
| Subject 1 (N = 13) | 28.15 | 4.18 | <0.01 | 28.54 | 4.16 | <0.01 | 28.31 | 3.97 | <0.01 |
| Subject 2 (N = 16) | 30.25 | 3.33 | <0.01 | 30.63 | 3.51 | <0.01 | 30.56 | 3.26 | <0.01 |
| Subject 3 (N = 19) | 32.37 | 2.71 | <0.01 | 30.89 | 3.02 | <0.01 | 33 | 2.62 | <0.01 |
| Subject 4 (N = 10) | 27.2 | 2.44 | <0.01 | 27.5 | 2.55 | <0.01 | 27.5 | 2.22 | <0.01 |
| Subject 5 (N = 13) | 27.77 | 2.08 | <0.01 | 28.38 | 2.39 | <0.01 | 28 | 2.08 | <0.01 |

### TABLE 8.3
### Paired t-test Applied on Circumferential Values before Modification

| Before Modification | Residual limb-BioSculptor | | | BioSculptor Model-Conventional | | |
|---|---|---|---|---|---|---|
| | Mean Dif. | SD | Sig. (2-tailed) | Mean Dif. | SD | Sig. (2-tailed) |
| Subject 1 (N = 13) | −0.385 | 0.650 | 0.054 | −0.054 | 0.376 | 0.165 |
| Subject 2 (N = 16) | −0.375 | 0.500 | 0.009 | −0.313 | 0.479 | 0.020 |
| Subject 3 (N = 19) | −0.526 | 0.612 | 0.001 | −0.632 | 0.596 | 0.000 |
| Subject 4 (N = 10) | −0.300 | 0.483 | 0.081 | −0.300 | 0.483 | 0.081 |
| Subject 5 (N = 13) | −0.615 | 0.506 | 0.001 | −0.231 | 0.439 | 0.082 |

objective. The values obtained shows the mean difference value between BioSculptor and POP models with reference to residual limb values.

One sample t-test is used for sample size to measures the size of the difference relative to the variation in sample data. The maximum SD is in case of Subject 1, where the dispersion is 4.027 and the minimum value dispersion on both sides is in case of Subject 5, 2.150. both these values are in case of residual limb readings while dispersion is not widely distributed in case of POP and BS readings in Table 8.2. Paired sample t-test shows that the p-value of the residual limb with BioSculptor is more significant than the conventional model in this case shown in Table 8.3. The dispersion in values is due to the nature of residual limb. Depending on muscle bulk and pressure application there is an alteration in few readings. However, the readings are not much different except for few where the application of pressure is in play. It is being noted that due to expanding property of POP and contours not absolute, there is more diversion while readings of BS are closer to natural. At the end, after modification and reduction as required values are altered.

### 8.4.2 PEARSON CORRELATION COEFFICIENT

Pearson correlation coefficient was applied on values after modification to measure the coalition profile difference at each interval to correspond the second objective of this study. The values obtained from the residual limb, POP and BS are analyzed using Pearson correlation coefficient. The bivariate correlations were measured twice for each subject. Initially linear relationship was checked between residual limb and BS and then between residual limb and POP for every result obtained for the final models.

Analysis was performed using SPSS on standard deviations and significant values of the BS and POP models comparing with the residual limb values as shown in Table 8.4.

Applying one sample statistics for single pair test to find the significance, mean, standard deviation and confidence interval it was found that the values are significant. The data is normally distributed which was checked using Shapiro–Wilk test

---

### TABLE 8.4
### The Standard Deviation and Significance of Circumferential Values of the Residual Limb vs. BS and POP after Modifications Using One Sample t-test

| After Modification | Residual Limb | | | BioSculptor Model | | | POP Model | | |
|---|---|---|---|---|---|---|---|---|---|
| | Mean | SD | P-Value | Mean | SD | P-Value | Mean | SD | P-Value |
| Subject 1 (N = 13) | 28.15 | 4.18 | <0.01 | 27.92 | 4.13 | <0.01 | 28.08 | 3.93 | <0.01 |
| Subject 2 (N = 16) | 30.25 | 3.33 | <0.01 | 30.38 | 3.59 | <0.01 | 30.31 | 3.36 | <0.01 |
| Subject 3 (N = 19) | 32.37 | 2.71 | <0.01 | 32.63 | 2.83 | <0.01 | 32.63 | 2.71 | <0.01 |
| Subject 4 (N = 10) | 27.2 | 2.44 | <0.01 | 27.4 | 2.59 | <0.01 | 27.3 | 2.36 | <0.01 |
| Subject 5 (N = 13) | 27.77 | 2.08 | <0.01 | 28 | 2.34 | <0.01 | 27.85 | 2.115 | <0.01 |

## TABLE 8.5
## Paired t-test Applied on Values before Modification

| Before Modification | Residual Limb-BioSculptor | | | BioSculptor Model-Conventional | | |
|---|---|---|---|---|---|---|
| | Mean Dif. | P-Value | Sig. (2-tailed) | Mean Dif. | P-Value | Sig. (2-tailed) |
| Subject 1 (N = 13) | 0.231 | 0.439 | 0.082 | 0.077 | 0.641 | 0.673 |
| Subject 2 (N = 16) | −0.125 | 0.500 | 0.333 | −0.063 | 0.574 | 0.669 |
| Subject 3 (N = 19) | −0.263 | 0.452 | 0.021 | −0.263 | 0.452 | 0.021 |
| Subject 4 (N = 10) | −0.200 | 0.632 | 0.343 | −0.100 | 0.316 | 0.343 |
| Subject 5 (N = 13) | −0.231 | 0.439 | 0.82 | −0.077 | 0.277 | 0.337 |

for normality. Standard deviation among values differs from patient to patient as represented in the table. Table 8.5 indicates that paired t-test was applied to further investigate the details of the accuracy in the circumferential profile of both BioSculptor and conventional models with reference to residual limb to correspond the first objective. The values obtained shows the mean difference value between BioSculptor and POP models with reference to residual limb values.

One sample t-test is used for sample size to measures the size of the difference relative to the variation in sample data. The maximum SD is in case of Subject 1, where the dispersion is 4.027 and the minimum value dispersion on both sides is in case of Subject 5, 2.150. both these values are in case of residual limb readings while dispersion is not widely distributed in case of POP and BS readings in Table 8.4. Paired sample t-test shows that the p-value of the residual limb with BioSculptor is more significant except for subject 5 than the conventional model in this case shown in Table 8.5. The dispersion in values is due to the nature of residual limb. Depending on muscle bulk and pressure application there is an alteration in few readings. However, the readings are not much different except for few where the application of pressure is in play. It is being noted that due to expanding property of pop and contours not absolute, there is more diversion while readings of BS are closer to natural (Karni and Karni, 1995).

### 8.4.3 VOLUMETRIC ANALYSIS

Volumetric analysis was done initially for one subject where the different procedures were applied and explained as follows:

#### 8.4.3.1 Water Displacement Method

*Residual limb volume measurement:* Residual limb is placed into a bucket of water slowly until the Supra Condylar wedge level, water can flow out from the bucket until it stops by itself. Water displaced is measured using measuring cup to obtain volume in cubic centimetres (cm³). This procedure is repeated ten times to get the average value (Figure 8.15).

Socket volume measurement: The internal volume of both sockets is determined by filling the socket up to the SC wedge depth with water. The volume in cubic centimetres is calculated by transferring the amount of water displaced into a measuring cup (cm³). To obtain the average value, this process is repeated ten times (Figures 8.16, 8.17).

**FIGURE 8.15** Measuring residual limb volume using water displacement method.

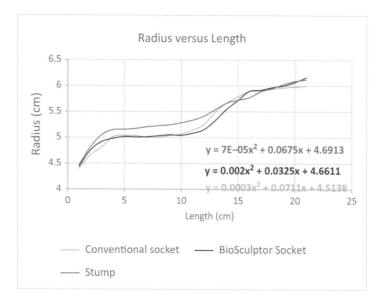

**FIGURE 8.16** Radius at 1cm intervals of duration for residual limb, inner radius for traditional socket and BioSculptor-fabricated socket, and equations for each graph line.

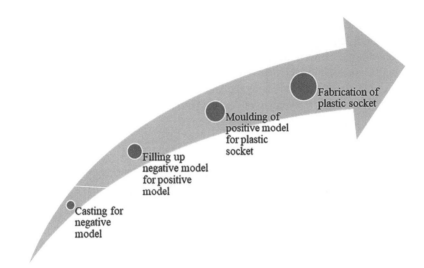

**FIGURE 8.17**  An illustration of the standard way of creating a transtibial socket.

Tables 8.6 and 8.7 indicate the percentage of inner volume gap between the BioSculptor-fabricated socket and the traditional socket in comparison to the residual arm. Negative percentage values indicate that the data is on the correct track. This is because each socket must be adjusted to accommodate the residual limb. Hence, the inner volume of both sockets must be smaller than the volume of residual limb. The four techniques used to determine residual limb volume yielded varying percentage differences. The BioSculptor socket has a greater gap (−7.4%) than the traditional socket by using the water displacement process (−6.09%). The disc model process yielded a close percentage gap for both sockets, about −3.6%. The Frustum sign model approach revealed that the traditional socket yielded a higher percentage of −3.5% than the BioSculptor-fabricated socket, which yielded a percentage of −1.5%. About the fact that this value is minimal, it seems to be an outlier in this data set. This is because the margin of error for all three indirect approaches ranges from

**TABLE 8.6**

**Using the Water Displacement Form, the Frustum Sign Model and the Integral of the Graph, Calculate the Percentage of Inner Volume Differential between the BioSculptor-Fabricated Socket and the Residual Limb**

| | Volume (cm³) | | |
| | | BioSculptor- | |
| Method | Residual Limb | Fabricated Socket | Difference (%) |
| --- | --- | --- | --- |
| Water displacement | 1916.67 | 1,766.43 | −7.84 |
| Disc model | 1967.61 | 1,897.45 | −3.57 |
| Frustum sign model | 1,865.35 | 1,837.25 | −1.51 |
| Integral of graph | 1,871.79 | 1,795.35 | −4.08 |

TABLE 8.7

The Percentage of the Inner Volume Differential between a Traditional Socket and a Residual Limb Calculated Using the Water Displacement Form, the Frustum Sign Model and the Integral of the Graph

| Method | Volume (cm³) | | Difference (%) |
|---|---|---|---|
| | Residual Limb | Conventional Socket | |
| Water displacement | 1916.67 | 1,800.00 | −6.09 |
| Disc model | 1967.61 | 1,896.27 | −3.63 |
| Frustum sign model | 1,865.35 | 1,799.12 | −3.55 |
| Integral of graph | 1,871.79 | 1,805.02 | −3.57 |

−3% to −4.1%. According to the findings of the analysis, there is a strong association between the water displacement method and the disc model approach, while the Frustum sign model is inapplicable. When the percentage of volume differential was calculated using the integral of graph equation, the BioSculptor-fabricated socket made a larger percentage difference (−4.1%) than the traditional socket (−3.6%).

This research demonstrates the existence of several tools for calculating the number of residual limbs and sockets. The disc model method and the integral of graph method to achieve volume are valid based on the percentage of variance for each method. Whereas more research into the implementation of the Frustum sign model system is needed. In this analysis, the water displacement procedure for measuring residual limb volume is deemed to be the most reliable. Yet, since the subject has lost all upper limbs, the measurement technique has become more complex and inaccurate because the subject is unable to balance while standing. Furthermore, the flow of water out of the bucket took a span of time that increased the subject's difficulties in stabilizing with one flexed leg standing while the subject lowered his remaining limb into the water bucket. The results of the circumferential profile and volumetric analysis in this thesis demonstrated that BioSculptor technology can be used in place of the traditional way of fabricating a transtibial socket.

However, technology can only function efficiently if the individual is knowledgeable in prosthetics and orthotics. Though the scanning of the residual limb is simple with a bioscanner, the alteration procedure with BioShape Manipulation tools necessitates technical comprehension as well as realistic practise to complete correct modifications on the digital image.

It is usually found that the BS illustrates higher value at proximal part of the prosthesis while the POP displays higher values at the distal end when compared to each other. The results and findings in this study reflects that BioSculptor technology can help in production of transtibial sockets more precisely and can be adopted for fabricating transtibial sockets which was also found by (Cheah et al., 2003). To be familiar with BioSculptor one must take training and should have basic knowledge of prosthetics and orthotics field. The scanning method is very simple but working on BioShape to modify and do the necessary changes in the model requires understanding about the theoretical application and practical experience for proper modification to obtain a good model which replicates the residual limb. Both methods have advantages depending on different circumstances. Sometimes for the patient, when we cannot use POP bandage to avoid infection on lesion or burn injury patients, the scanning technique is preferable. Conventional techniques give more control of shaping the socket than scanning (Herbert et al., 2005). The scanning technique can help preserve the model, to restore it if there is any problem in the process, without needing the patient to visit again. The conventional models occupy a lot of space in the workshop, while the scanning technique stores models digitally (Singh, 2013). BioSculptor technology can help in reducing cost and time in fabricating sockets. It also gives the advantage of using the digital image. It also helps in getting unlimited scans of the scan unlike manual casting. Modification in BioShape software gives us the edge to modify the model and undo the changes which is not possible in manual modification of the POP model. POP takes a long time to discompose, and it is not environmentally friendly so BioSculptor technology can help us in replacing the old technique because of its advantages over conventional manual technique. Both techniques are significant where p<0.001 for both methods using t-test with 95% of confidential interval where BS shows less standard deviation than the values obtained from POP models. The average t-value of BS is 38.86 with reference to the residual limb values 39.48 while for POP is 40.40. Pearson correlation coefficient was applied to the values against the residual limb values, and it was found that both BS and POP are significant with P<0.001 giving us a straight line in the graph. The average $R^2$ for BS is 0.999 while for POP is 0.998, where BS shows stronger relationship compared to the POP values. There were differences at different levels and both techniques give good result but overall BioSculptor gives accurate values then the manual technique.

## 8.5  CONCLUSION

Circumferential profile comparison shows that BioSculptor technology provides precise values compared to the conventional manual casting method after modification. In BioSculptor, the exact amount of modification can be done digitally in the system, while in manual modification of the POP model there is no precision involved other than experience and skills. It was found that the POP model expands while filling the negative casts which cannot occurs in BioSculptor after scanning. The comparison also shows that scanning technique takes very less time compared to the manual casting method. This helps us in fabricating more sockets in less time. If comparing the cost of the socket fabrication, the BioSculptor seems to be cheaper than the manual technique. The advantage of the BioSculptor is that if any problem occurs during the modification, it can use the saved model, which is not possible in manual methods; when any problem occurs, we must call the amputee in again to take another impression by casting using POP bandage. BioSculptor stores the file in the computer unlike manual method where the models are kept in the workshop and sometimes it is difficult to accommodate more models due to less space available in the workshop. Another advantage of the BioSculptor is that the scan does not trouble the patients with sensitive residual limb, whereas using POP bandage process can make the patient feel uncomfortable. One advantage of the BioSculptor is that it does not make the place dirty and messy unlike in casting using POP powder and then modification of the model using surform blade. The structural feeling of the residual limb using POP bandage for taking impression of the residual limb and can capture the shape nicely which is not possible in scanning using BioSculptor. A comparison of circumferential profiles revealed that a socket made with BioSculptor technology is interchangeable with a traditional socket with greater precision. As a result, BioSculptor technology is similar to the traditional approach.

It is a procedure that can be used to make a prosthetic transtibial socket if the user has experience of the prosthetic and orthotic fields. Both the techniques can be used depending on the facility and location of the workshop. BioSculptor technology uses engineering knowledge for understanding the software while the manual method requires clinical skills. The BioSculptor solution system is very expensive but is a good tool for research purpose as well as industrial use. It was found that BioSculptor gives more precise modification digitally compared to manual modifications, which can help in producing a well-fitted socket without discomfort. As transtibial residual limbs are one of the complicated residual limbs so it is difficult to make a comfortable socket. Thus, this technology can help in producing sockets following the shape and contours of the residual limb for making a well-contoured socket. The coalition profile was measured at different layers on the residual limb, and it was found that BioSculptor gives more accurate results compared to the conventional system. This will help us getting precise values at every level because most of the time, patients with redundant tissue on the distal end feel pain during fitting. It also helps in getting the pressure distributed evenly across the residual limb. The results were significant in both techniques, but the BioSculptor gives accurate values which was closer to the residual limb values of the amputee.

## 8.6 STUDY LIMITATION AND FUTURE DIRECTION

There were some limitations in the study. It was difficult to get used to the software initially, as there was no training, so the researcher had to learn and practice, taking time to get familiar with the software for fabricating transtibial sockets. The initial few trials were not successful as good shapes were not reproduced of the residual limb, because the researcher was not familiar with the software. Later, with practice, better results were obtained. There were occasional technical issues with the software because of licence expiry, and there were delays in new licencing generation. Once the template was formed for a specific model, it was easier to cope on subsequent models following similar methods. There were limitations of budget for the research causing the researcher to change parameters on some occasions. In future, this research should continue to have fitting of sockets with Tek-scans to check the interface pressures and get the feedback from the subjects in a blind study to find out which socket is more comfortable and exerting less forces and pressures on the residual limb. These experiments will help us determine more precisely about which fabrication technique helps in producing well-contoured and comfortable sockets. Gait analysis should be performed to see the pressures at different phases of the gait cycle. In future this research can be carried further to validate the conclusions provided more nicely and precisely. This research will be be more significant in clinical and industrial use for fabricating well-fitted sockets without discomfort.

## REFERENCES

BioSculptor Innovative Solutions. (1997b). Bioscanner. Retrieved from http://BioSculptor. com/bioscanner/

BioSculptor Innovative Solutions. (1997c). BioSculptor Innovative Solutions. Retrieved from http://BioSculptor.com/

Bolt, A., de Boer-Wilzing, V., Geertzen, J., Emmelot, C., Baars, E., & Dijkstra, P. (2010). Variation in measurements of transtibial residual limb model volume: a comparison of five methods. *American Journal of Physical Medicine & Rehabilitation*, *89*(5), 376–384.

Chan, J. C., Malik, V., Jia, W., Kadowaki, T., Yajnik, C. S., Yoon, K.-H., & Hu, F. B. (2009). Diabetes in Asia: epidemiology, risk factors, and pathophysiology. *Jama*, *301*(20), 2,129–2,140.

Cheah, C. M., Chua, C. K., & Tan, K. H. (2003). Integration of laser surface digitizing with CAD/CAM techniques for developing facial prostheses. Part 2: Development of molding techniques for casting prosthetic parts. *International Journal of Prosthodontics*, *16*(5), 543–548.

Childress, D. S. (2002). Presentation highlights: computer-aided design and manufacture (CAD-CAM). *Journal of Rehabilitation Research and Development*, *15*. Washington.

Chow, D. H., Holmes, A. D., Lee, C. K., & Sin, S. (2006). The effect of prosthesis alignment on the symmetry of gait in subjects with unilateral transtibial amputation. *Prosthetics and Orthotics International*, *30*(2), 114–128.

Courvoisier, A., Sailhan, F., Thevenin-Lemoine, C., Vialle, R., & Damsin, J.-P. (2009). Congenital tibial deficiencies: treatment using the Ilizarov's external fixator. *Orthopaedics & Traumatology: Surgery & Research*, *95*(6), 431–436.

Dillingham, T. R., Pezzin, L. E., & MacKenzie, E. J. (2002). Limb amputation and limb deficiency: epidemiology and recent trends in the United States. *Southern Medical Journal*, *95*(8), 875–883.

Goh, J. C. H., Lee, P. V. S., & Chong, S. Y. (2004). Comparative study between patellar-tendon-bearing and pressure cast prosthetic sockets. *Journal of Rehabilitation Research & Development*, *41*, 491–501.

Herbert, N., Simpson, D., Spence, W. D., & Ion, W. (2005). A preliminary investigation into the development of 3-D printing of prosthetic sockets. *Journal of Rehabilitation Research and Development*, *42*(2), 141.

Jia, X., Zhang, M., & Lee, W. C. (2004). Load transfer mechanics between transtibial prosthetic socket and residual limb—dynamic effects. *Journal of Biomechanics*, *37*(9), 1,371–1,377.

Johansson, S., & Öberg, T. (1998). Accuracy and precision of volumetric determinations using two commercial CAD systems for prosthetics: a technical note. *Journal of Rehabilitation Research and Development*, *35*(1), 27.

Karni, J. & Karni, E. Y. (1995). Gypsum in construction: origin and properties. *Materials and Structures*, *28*(2), 92–100.

Michael, J. W., & Bowker, J. H. (Eds.). (2004). *Atlas of amputations and limb deficiencies: surgical, prosthetic, and rehabilitation principles*. Rosemont, IL: American Academy of Orthopaedic Surgeons.

Mohamud, W. N. W., Musa, K. I., Khir, A. S. M., Ismail, A. a.-S., Ismail, I. S., Kadir, K. A., Kamaruddin, N. A., Yaacob, N. A., Mustafa, N., Ali, O. and Md Isa, S. H., Ali, O. (2011). Prevalence of overweight and obesity among adult Malaysians: an update. *Asia Pacific Journal of Clinical Nutrition*, *20*(1), 35–41.

Öberg, T., Lilja, M., Johansson, T., & Karsznia, A. (1993). Clinical evaluation of transtibial prosthesis sockets: a comparison between CAD CAM and conventionally produced sockets. *Prosthetics and Orthotics International*, *17*(3), 164–171.

Organization, W. H. (2004). *The world health report: 2004: changing history.*

Radcliffe, C. W. (1961). *The patellar-tendon-bearing below-knee prosthesis.* Tokyo, Japan: Biomechanics Laboratory.

Robert Gailey, P. (2008). Review of secondary physical conditions associated with lower-limb amputation and long-term prosthesis use. *Journal of Rehabilitation Research and Development, 45*(1), 15.

Sanders, J., Zachariah, S., Baker, A., Greve, J., & Clinton, C. (2000). Effects of changes in cadence, prosthetic componentry, and time on interface pressures and shear stresses of three transtibial amputees. *Clinical Biomechanics, 15*(9), 684–694.

Singh, U. (2013). Role of cad-cam technology in prosthetics and orthotics. *Essentials of Prosthetics and Orthotics, 86*(1).

Staats, T. B., & Lundt, J. (1987). The UCLA total surface bearing suction below-knee prosthesis. *Clinical Prosthetics and Orthotics, 11*(2). 118–130.

Steen Jensen, J., Poetsma, P., & Thanh, N. (2005). Sand-casting technique for transtibial prostheses. *Prosthetics and Orthotics International, 29*(2), 165–175.

Vamos, E. P., Bottle, A., Edmonds, M. E., Valabhji, J., Majeed, A., & Millett, C. (2010). Changes in the incidence of lower extremity amputations in individuals with and without diabetes in England between 2004 and 2008. *Diabetes Care, 33*(12), 2,592–2,597.

Wolf, S. I., Alimusaj, M., Fradet, L., Siegel, J., & Braatz, F. (2009). Pressure characteristics at the residual limb/socket interface in transtibial amputees using an adaptive prosthetic foot. *Clinical Biomechanics, 24*(10), 860–865.

Wu, Y., Casanova, H., Smith, W., Edwards, M., & Childress, D. (2003). CIR sand casting system for transtibial socket. *Prosthetics and Orthotics International, 27*(2), 146–152.

Yiğiter, K., șener, G., & Bayar, K. (2002). Comparison of the effects of patellar tendon bearing and total surface bearing sockets on prosthetic fitting and rehabilitation. *Prosthetics and Orthotics International, 26*(3), 206–212.

Ziegler-Graham, K., MacKenzie, E. J., Ephraim, P. L., Travison, T. G., & Brookmeyer, R. (2008). Estimating the prevalence of limb loss in the United States: 2005 to 2050. *Archives of Physical Medicine and Rehabilitation, 89*(3), 422–429.

# 9 Biomechanics of Posture and Stance-Phase Stability

*N A Abu Osman*

University of Malaya, Kuala Lumpur, Malaysia

*P X Ku*

School of Computer Science & Engineering, Subang Jaya, Malaysia

## CONTENTS

## 9.1 INTRODUCTION

Postural balance is an important skill to maintain our regular daily activities. A proper postural balance involves different locomotion types such as standing, walking, running and sitting. Amongst all movements, upright standing is the essential locomotion posture for humans. A toddler starts applying motor skills by crawling on the ground. In the next stage, the toddler learns to control his trunk movement to

DOI: 10.1201/9781003196730-9

stand still. If he can stand independently and confidently, he will try to perform walking, running or other types of locomotion.

Loss of balance may lead to critical general health problems such as strain in the tendon or joint structure, or increase the risk of fall-related injuries (Ludwig et al., 2020). As the second-largest public health problem, falls are the most prominent external condition of unintentional injuries. It is defined as the inadvertent action of go down onto the ground and typically happens spontaneously (World Health Organization, 2008). If an individual could not recover his postural balance in response to the perturbation, he may fall and be injured. In the US, around 36 million people experience a fall each year (Moreland et al., 2020). According to the injury statistics by the Centers for Disease Control and Prevention, it was reported that fall-related injury among people aged 65 and over was experienced about three times more frequently than other age groups (Table 9.1).

The risk of bone fracture was associated with anthropometric characteristics among the normal elderly population (Compston et al., 2014). The health condition for 30% of elderly aged 75 years and over are in poor health; while 10% of the elderly may experience moderate accidental injuries such as fracture, joint dislocation, severe head passive or mortality (Marchetti et al., 2011). An estimate of 58% of trauma admissions in the elderly occurs through standing height falling and nearly 30–50% of fall-related injuries are induced by involuntary tripping and slipping (Hall et al., 2019). Moreover, illness or diseases such as osteoporosis, stroke or post-fall syndrome may adversely affect postural stability and lead to an increase in falling risk.

Studies in human biomechanics are essential to provide fundamental knowledge on human mobility related to public health. In human biomechanics, postural balance refers to the fundamental skill to regulate our mobility or locomotion. By regulating the body segment, the centre of mass is orienting within the base of support to achieve the equilibrium against balance disturbance (Alexandrov et al., 2005). Kinetic and kinematic data are considered to fully describe human standing posture. The static stance posture is the body segments aligned upright to archive static equilibrium

**TABLE 9.1**

**The Total Cases of Non-Fatal Injuries Due to Falling during the Year 2019**

| Age Group | Total Non-fatal Injuries (%) |
|---|---|
| < 15 years | 2.50 |
| 15 to 19 years | 1.39 |
| 20 to 44 years | 1.27 |
| 45 to 64 years | 2.05 |
| 65 to 74 years | 3.50 |
| < 75 and over | 9.05 |

(Adapted from the National Center for Injury Prevention and Control, 2019. CDC WISQARS™ — Your source for US Injury Statistics, October 25, 2019 ed. Centers for Disease Control and Prevention, National Center for Injury Prevention and Control)

when someone standing still. In human biomechanics, the term "centre of mass" (CoM) refers to a position on the body that all the body mass is considered, where the average mass distribution of all body segments is assumed to act against balance disturbance (Hernández et al., 2009). For the term "base of support" (BoS), it is defined as the displacement area of the centre-of-pressure that the net ground reaction force acts against the supporting surface, and the centre-of-pressure is restrained to act within it (Barrett et al., 2012). The term "centre-of-pressure" (CoP) is defined as the standard deviation of the instantaneous position that the resultant of all ground reactions forces acts under the feet (Collins and Luca, 1993). The trajectory of CoP is commonly used as the measured parameter in human biomechanics studies. The position of the earlobe, cervical vertebrae, shoulder, thorax, lumbar vertebrae, posterior to the hip joint, anterior to the knee joint, and lateral malleolus formed the line of gravity for a normal posture alignment (Kisner and Colby, 2002). Postural sway is quantified by measuring the CoP motion on the contact surface. The increase of postural sway required more effort in the control of the body trunk to achieve body equilibrium.

## 9.2    FACTORS CONTRIBUTING TO THE NATURE OF POSTURAL CONTROL MECHANISM

The postural balance involves various complex systems such as the human central nervous system, sensory system and musculoskeletal system. It may also be affected by external effectuating factors such as body fatigue, sleepiness, hormone levels, or anthropometric characteristic (Jorgensen et al., 2012). However, the alterations in any of the systems or external effectuating factors may adversely impact postural control. There are six main conditions that affect the postural balance of an individual: sensory strategies, movement strategies, biomechanical task constraint, controls of dynamic, orientation in space, and cognitive processing (Horak, 2006) (Figure 9.1).

### 9.2.1   SENSORY STRATEGIES

The inputs of the physiological system are important to sustain body equilibrium. It involves the visual system, vestibular system and proprioception system. The visual system is responsible for detecting information from the surrounding environment and transmitting the data to the primary visual cortex for further visual processing. Generally, the visual input is mainly involved in the balance control for low-frequency disturbance and movements (Gill et al., 2001; Tomomitsu et al., 2013). It also provides information related to physical and virtual distance estimation in body adjustment against perturbation. The eyes-closed condition tends to increase the CoP displacement and velocity (Yoon et al., 2012). Besides, Hafstrom et al. (2002) revealed that the stance balance with the eyes-open and eyes-closed condition is identical in a dark environment. It was speculated that a dark environment may cause a visual feedback transmission delay in the eyes-open condition.

The vestibular system is a sensor motor system that is responsible to regulate the sense of balance and movement in our daily life. The vestibular input helps to detect the head position in space, rotary head movement, stabilizes the eyes during head motion and helps in motor response coordination (Netter, 2010). The common cause

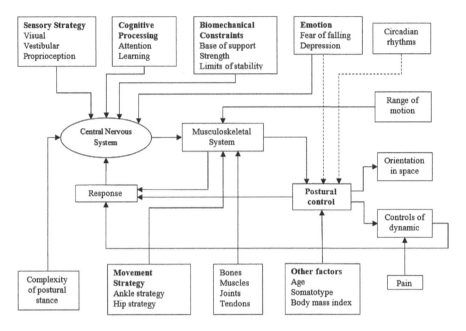

**FIGURE 9.1** The interaction of various components related to postural stance control. The dashed line refers to the indirect effect of emotion on physical activities.

of vestibular dysfunction includes ageing, traumatic brain injury, or infections (Marchetti et al., 2011). The compensation of vestibular deficit could reduce the fear and risk of falling, and improve the postural balance of the elderly.

For the proprioception system, it is a sensitivity mechanism that transmits information related to the strength of effort and changes of location that is applied in the movement to the central nervous system for information processing (Shier et al., 2008). The terms "sensation" and "perception" are often used to describe the types of proprioceptive information perceived by the sensory organs from the surrounding environment. The sensation is referred to as the physiological process that involves sensory organs, receptors, neural pathways and the central nervous system; while perception is defined as the psychological process that involves derivation and interpretation of the sensation input (Drake et al., 2009). The sensation input involves regulating the movement with the help of the spinal cord in the movement control. With the help of contrast sensitivity, it allows us to detect the edge and shape of any object (Seeley et al., 2006). Besides regulating the dynamic restraints, the proprioception system could affect muscle stiffness and dynamic joint stability (Riemann and Lephart, 2002). The proprioception receptor located in joints, tendons, muscles and feet is used to identify the changes of muscle force, joint position, limb orientation, and foot pressure distribution in stance posture (Mehdikhani et al., 2014). The ankle proprioception mainly helps to adjust the ankle position and movement of the upper body in complex motor tasks (Han et al., 2015).

As a healthy individual, vision (10%), vestibular (20%), and proprioceptive (70%) information are used to preserve balance control under a firm BoS (Horak, 2006).

The sensory system relies on both the visual and vestibular input and reduces the need for proprioceptive inputs during an unstable supporting surface condition (Peterka, 2002). Besides, the individual with proprioception loss or peripheral vestibular loss from neuropathy has experienced a limitation in postural sensory dependence, further increase the risks of falling (Jáuregui-Renaud, 2013). Other than that, the sensory input response would deteriorate when we are getting older.

## 9.2.2 MOVEMENT STRATEGIES

Postural balance is generally related to the body segment movement. By constantly orientating our body segments, the body equilibrium is achieved with different postures in the activities such as standing, walking, and sitting (Hernández et al., 2009; Del Porto et al., 2012). The three balance recovery strategies – ankle, hip, and step – are used mainly to regain equilibrium against perturbation (Winter, 1995). Ankle strategy is applied in response to a small perturbation forces, whilst step strategy is applied in response to a much larger perturbation. Both the ankle and hip strategies require an intact range of motions and forces in the ankle and hip joint, whereas the step strategy is used to realign the BoS and CoM. Furthermore, the contact forces, joint torque, muscle synergies, and movement patterns are involved in regulating the CoM location on the sagittal plane (Horak et al., 1997). The changes of weight-bearing between two extremities will generate sufficient joint torques to stabilize the body trunk.

The musculoskeletal system serves as an important component in regulating body movement. The deterioration in muscle tone strength and motor skill will diminish the postural stability (de Oliveira et al., 2008). The muscles involved in balance recovery strategies are gastrocnemius, tibialis anterior, hamstring, quadriceps femoris, paraspinal and abdominal muscles. The gastrocnemius muscle is the largest calf muscle situated in the rear of the lower limb, while the tibialis anterior muscle is located on the lateral surface of the tibia. The hamstring muscles are located at the rear of upper limb. It consists of three posterior thigh muscles: biceps femoris, semitendinosus and semimembranosus. For the quadriceps femoris muscles, it refers to four muscles: vastus lateralis, vastus medialis, vastus intermedius and rectus femoris. These are situated at the front of the upper limb. The paraspinal muscles are located parallel to the spine. It helps to control the angular motion of the spine and body trunk, to avoid hyperextension spine injury. The abdominal muscle is located at the mid-section of the abdomen. Muscle synergies helps to coordinate movement as all the muscle groups act and work together as a unit in the motor control.

The ankle strategy is applied if the line of gravity shifts anteriorly to the knee joint when a small balance disturbance occurs. The adjustment of ankle strategy mainly relies on the proprioceptive inputs and could also be impacted by the BoS area (Horak and Nashner, 1986). In this strategy, the muscles acting on the ankle joint are activated and induced to generate the ankle torques (Winter, 1995). To reverse the forward body motion, the gastrocnemius muscle exerts the plantar-flexion torque, while the activation of paraspinal and hamstring muscles affects the proximal body segments by regulating the knee and hip joint (Kuo and Zajac, 1993). The erector spinae muscles limit excessive angular motion of trunk movement and cause the body trunk to sway backward (Horak and Nashner, 1986; Shumway-Cook and Woollacott, 2007). Apart from the mentioned muscle groups, activation of iliopsoas muscle helps to restrict the occurrence of hip hyperextension.

The hip strategy allows the activation of hip and trunk muscles and exerts large and rapid motion of the ankle, knee, and hip joint torques (Winter, 1995). The adjustment of hip strategy mainly relies on vestibular inputs. The hip strategy is applied when a larger perturbation occurs in the forward direction. The activation of the quadriceps femoris and abdominal muscles generates the backward motion of the body trunk to restore the body equilibrium (Salsabili et al., 2011).

The ankle joint is important in the control body sway for the anteroposterior direction. The ankle response is responsible to maintain the body upright in the quiet stance. Whereas hip joints maintain postural stability by regulating the body sway in the mediolateral direction. The hip response is mainly used for balance recovery in a larger CoM perturbation (Salsabili et al., 2011). It appears that the ankle response will compensate for the balance deficiency results from the small external perturbation in the ankle strategy. However, the ankle response is insufficient to regain balance in a larger disturbance. Therefore, the hip response is used in response to a larger balance perturbation rather than rely on the ankle response (Park et al., 2004)

### 9.2.3 BIOMECHANICAL CONSTRAINTS

The biomechanical constraint or whole-body mechanical consideration is the biomechanical factors such as or limits of stability, strength, degree of freedom, that could directly or indirectly affect the body equilibrium. The most important biomechanical factor used in postural control is the quality and size of the BoS (Tinetti et al., 1988). The area of the body segments (e.g., hands or feet) on the contact surface is defined as the BoS area (Figure 9.2). In a balance disturbance, the CoM is much easier to be relocated back within the wider BoS area than a narrow area.

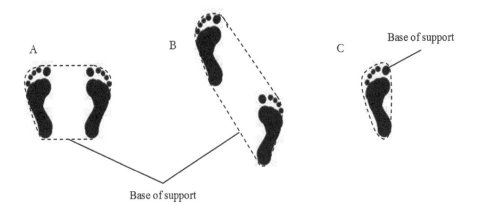

**FIGURE 9.2** The surface area of the BoS in different stance conditions. (A) bipedal stance, (B) split stance, (C) unipedal stance.

Moreover, individual with smaller limits of stability and the BoS tends to suffer a higher fall risk. The postural control system will automatically adjust according to the task requirement to ensure the CoM is being relocated with the BoS. If the line of gravity shifts out from the BoS, a torque related to the body angular motion generates and deteriorates the body equilibrium.

Additionally, higher-level of mobility skill are greatly influenced by the quality of the BoS and CoP location. A small BoS area restricts the allowable CoM motion in the transverse plane and the control of mediolateral stability (Mehdikhani et al., 2014). However, the extensive BoS area could restrict ankle mobility, postural sway frequency, and improve biomechanical stiffness in the mediolateral direction (Winter, 1995). The CoM height is relatively associated with the BoS in response to the postural control. If the position of CoM is located outside the BoS, the body will experience balance deficits for dynamic postural control (Shumway-Cook and Woollacott, 2007). Besides, the width of foot placement or area of BoS should be equivalent to half of the shoulder width for a more stabilize posture in the quiet stance (Yoon et al., 2012).

### 9.2.4 Controls of Dynamic

The ability of dynamic balance control is another predominant element for gait and postural changes. Human "quiet stance" is a functional posture that involves the sensory inputs, joints, and body segments (Ku et al., 2012b). The balance control in gait and postural changes require a more complex system to maintain body equilibrium compared to quiet stance. The changes in weight distribution of the lower limb may push the CoM away from the BoS in the gait cycle. The CoM position of the swing phase controls the dynamic stability in the anteroposterior direction, whereas the lateral trunk controls the stability in the mediolateral direction (Horak, 2006). Teasdale and Simoneau (2001) found that a higher lateral excursion of CoM and asymmetry of lateral foot placement would increase the fall risk among the elderly population. The elderly tend to have a narrow step width and length to overcome obstacles, and rely more on central reorganization to sustain equilibrium (Woollacott and Tang, 1997).

### 9.2.5 Orientation in Space

The body movement acts upon the information related to perception, visual input, GRF, contact surface. The central nervous system is responsible to process and interpret the information obtained from the sensory systems. It will regulate the body segment to orient the CoM of the body segments, in order to restore the body to an equilibrium state. In the natural body alignment for the bipedal stance, the body trunk aligns perpendicularly over the flat supporting surface. In other words, the body will orient to adapt to the changes of the contact surface angle. The adjustment of gravitational vertical in the dark environment depends on the vertical postural perception (Bisdorff et al., 1996). In a dark environment, a healthy adult can tolerate no more than $0.5°$ of the gravitational vertical (Horak, 2006). Misinformation of internal representation of verticality decline the ability in postural alignment, further results body imbalance. The internal representation of visual verticality mainly affects the patient with unilateral vestibular loss. A stroke patient with hemispatial neglect may be affected by the internal representation of proprioception verticality.

## 9.2.6 COGNITIVE PROCESSING

Cognitive processing involved a person's attention and learning ability. The study by Mignardot et al. (2010) revealed that obese people require more cognitive engagement and attention to sustain body equilibrium than non-obese people in the unipedal stance. The learning ability is the ability to understand, apply and gain knowledge from experiences and practices. With the high level of neural cognitive processes, it helps to develop the anticipatory and adaptive abilities in motor control (Shumway-Cook and Woollacott, 2007). Moreover, a longer postural response and reaction time is required for the cognitive information processing in response to the complexity of the postural task (Teasdale and Simoneau, 2001).

## 9.2.7 OTHER RELATED FACTORS

### 9.2.7.1 Ageing

Previous studies consistently found that postural control deficit is related to ageing (Medina, 1996). The balance function and gait change and deteriorate accordingly when people are getting older. The age-related deterioration in the bone mass, muscle mass, strength and body motion will impair postural balance. Due to the ageing process, the decrease of muscle mass and bone mass in the lower extremities is greater compared to the upper extremities, further diminishes functional mobility and balance control. The elderly tend to suffer a higher hip fracture risk than the young population (Butler et al., 1996). Forty percent of lower limb muscle strength is diminished around the age of 30–80 years (Aniansson et al., 1986). The ageing process can also cause degeneration of the sensory system and delays the reaction time and postural response against balance disturbance. Thus, the elderly rely more on visual input in regulating balance control, whereas children rely on proprioceptive and baroreceptor inputs (Hytonen et al., 1993).

### 9.2.7.2 Body Mass Index and Somatotype

Body Mass Index (BMI) is a body fat estimation index based on the relative weight that applies to adults. It is calculated as body weight in kilograms divided by the square of body height in metres (Keys et al., 1972). There are four classifications in BMI: underweight (below 18.50 kg/m$^2$), normal weight (18.50–24.99 kg/m$^2$), overweight (25.00–29.99 kg/m$^2$), and obese (30.0 kg/m$^2$ and above). Since a relatively high BMI can be interpreted either as high body fat or lean body mass, BMI may only serve as an indicator of body fatness. Obesity is a medical condition related to the excessive amount of adipose tissue or body fat accumulation in the body that may adversely affect health conditions. It is generally associated with increased functional impairment, balance deficiency, and fall risks (Greve et al., 2007). The increased body mass impairs the functional ability and impose new biomechanical constraint to sustain body equilibrium (Hue et al., 2007; Blaszczyk et al., 2009).

Despite that, Cruz-Gómez et al. (2011) suggested the ageing factor may not affect the postural control with respect to the different BMI categories. Besides, the balance confidence level of obese people with mental disorders such as depression is indirectly affected (Klinitzke et al., 2012).

Somatotype is a classification of the human physique according to the shape of human body. There are three classifications of body types: ectomorphic (slim and tall), mesomorphic (strong and muscular), and endomorphic (fat or round) (Miller, 2014). The body shape normally results from the accumulation of adipose tissue in the different body parts. If the adipose tissue distributed around the hip and thigh region, it is known as gynoid type (pear-like body shape) fat distribution. The distributed adipose tissue in the thorax and abdomen region, is known as android type (apple-like body shape) fat distribution (Clark, 2004). Generally, these two types of fat distribution could be developed regardless of gender. Nevertheless, females are more likely to develop gynoid fat distribution, whereas the android fat distribution is more likely to happen in males. Most adipose tissue accumulation occurs at the lower extremities in obese females. Excessive adipose tissue would increase the difficulty to control the CoM and CoP within a given BoS.

### 9.2.7.3  Circadian Rhythms

Circadian rhythm is an internal process used to execute the essential functions that follow a 24-hour cycle. The postural control could be affected by the human biological rhythm, in response to the changes of environmental factors such as light, body temperature, metabolism and age. The postural control could be affected by the fluctuation of sleepiness level and body temperature and a long time period in a 24-hour cycle. A better postural could be found in the evening, with a lower sleepiness level and higher body temperature (Bougard and Davenne, 2014). Goel et al. (2013) conducted a limited/no sleep of psychomotor vigilance test. They found that sleepiness significantly affects the cognitive processing, attention demand and alertness and balance control of healthy adults.

## 9.3  POSTURAL CONTROL IN DIFFERENT STANCE CONDITIONS

Quiet stance is a basic functional and transition posture that relates to the line of gravity in response to the regulation of body segments to maintain an upright standing alignment. Bipedal stance and unipedal stance are the common stance postures studied in the balance control-related investigation. Bipedal stance is referred to standing with feet on a supporting surface, while unipedal stance is referred to standing with either dominant or non-dominant leg on a supporting surface. Unipedal stance is used to assess postural steadiness in quiet stance (Josson et al., 2004). Compared to bipedal stance, it requires greater motor control to maintain body equilibrium. The ankle and hip joint exert greater musculature restoration forces to maintain the postural equilibrium. Choy et al. (2003) found that elderly women aged 70 years suffer difficulty in static stance control.

To measure the balance ability, force platform is the measuring instrument used to assess the ground reaction force involves in human movement. The other instruments used to evaluate postural stance balance are the Zebris FDM-S system, Vicon system, NeuroCoM Balance Master and Biodex Balance System. The test-retest reliability of all mentioned instruments is acceptable for clinical testing. In this section, the postural control on static stance, dynamic stance and stance with toe extension will be explored and discussed.

### 9.3.1 STATIC STANCE POSTURAL CONTROL

In the static balance control, the analysis of postural sway changes is served as a balance predictor for dynamic control (Ku et al., 2014). The changes of postural control for static standing related to different measured parameters and conditions have been investigated. The increased body weight and BMI are associated with postural instability. Obese adults are more likely to suffer higher fall risk than non-obese adults (Corbeil et al., 2001). Hue et al. (2007) suggested that the increased pressure and contact area have diminished the sensory inputs detected by the plantar mechanoreceptors, further result in postural instability among middle-aged obese adults. In the study by Ku et al. (2012a), the changes between BMI and balance control in bipedal and unipedal stance conditions were assessed. It was found that obesity requires a greater postural sway to achieve body equilibrium in both the bipedal and unipedal stance (Figure 9.3). The postural sway is increased as a result of the inability of the obese group to exert sufficient muscle force to regulate CoM displacement in the quiet stance (Colne et al., 2008; Ku et al., 2012a). High BMI may cause inadequate

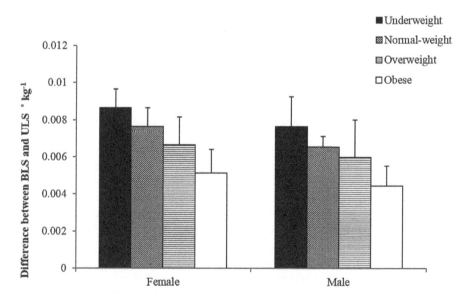

**FIGURE 9.3** The relative stability scores of males and females between bipedal stance and unipedal stance according to the BMI classification. (Reproduced from Ku, P.X., Abu Osman, N.A., Yusof, A., Wan Abas, W.A., 2012. Biomechanical evaluation of the relationship between postural control and body mass index. Journal of Biomechanics 45, 1,638–1,642.)

muscle force to regulate the CoM displacement. Meanwhile, leg dominance will not affect the postural control in unipedal stance condition (Greve et al., 2007). Moreover, people with a sedentary lifestyle are more likely to experience muscle mass reduction, body fat accumulation, weight gain and postural instability (Himes, 2000). If obese adults are willing to reduce their body weight, their postural control could be improved significantly (Teasdale et al., 2007).

In quiet standing, the increase of lateral body sway may cause postural instability. Studies found that changes of the mediolateral sway are greater than the anteroposterior sway, where people need to use more effort to regulate the lateral body sway (Blaszczyk et al., 2007; Ku et al., 2012a). Generally, increased body weight could affect the anteroposterior stability (Menegoni et al., 2009). The lateral balance control is achieved by regulating the motion of the pelvic against the balance perturbation. The ankle and knee joints are primarily used to control the anteroposterior stability, whereas the hip joint is used to control the mediolateral sway. The lower limb joints are more progressively misaligned with the line of gravity after the initiation of standing posture (Di Giulio and Baltzopoulos, 2019). The postural control for children, young adults and the elderly were investigated in many studies. All the findings were consistently shown that obese people tend to exert higher lateral sway compared to non-obese people in quiet stance (McClenaghan et al., 1996; McGraw et al., 2000; Ku et al., 2012a). For the unipedal standing, obese people also experienced a higher displacement in the mediolateral direction (Ku et al., 2012a). Furthermore, the postural control changes of anteroposterior sway for the young adult are not significant in the bipedal quiet standing. Nevertheless, the increased anteroposterior sway results in the deterioration of postural balance related to ageing. Menegoni et al. (2009) suggested that increasing motor activity is triggered to compensate for the anteroposterior instability.

The result about gender effect on postural control have been somewhat contradictory. There is a trend in the underweight and normal-weight females to exert greater postural sway compared to males in both bipedal stance and unipedal stance conditions (Ku et al., 2012a). This may result from the increase in CoP sway towards the BoS limit. Despite that, the finding of Mickle et al. (2010) reported that males tend to produce greater postural compared to females, due to the distribution of adipose tissue. Moreover, the increased posture sway in females may appear in conjunction with the arch angle of the foot. The arch angle exhibits a greater ligament laxity, and a flexible arch promotes greater postural displacement, further result in postural instability (Aurichio et al., 2011).

### 9.3.2   DYNAMIC STANCE POSTURAL CONTROL

Dynamic stance balance is the ability for the acquisition and execution of motor control through the regulation of centre of gravity, weight distribution, and muscle control in response to the changes of BoS (Sabashi et al., 2021). Limits of stability (LoS) is a type of dynamic standing task for fall risk screening, directional and functional stability evaluation (Horak et al., 2005). The LoS indicates the maximum voluntary distance or angle that a person can achieve by leaning the body within BoS towards specified targets. Unlike the static quiet stance, the body will lean outward from the upright position of quiet stance to reach the pre-specified target. The eight specified target directions are forward, forward-rightward, rightward, backward-rightward, backward, backward-leftward, leftward and forward-leftward directions. Reaching the targets while maintaining body equilibrium in the LoS test involves the integration of anatomical, functional, mechanical limit and internal body motion with the given BoS (Kolarova et al., 2013). A healthy person can easily control the directional stability in a static platform or contact surface. However, the increased tilt angle of the moving platform is associated with postural instability (Pugh et al., 2011). When the platform tilt angle changed from rest to 14°, the directional stability in rightward, backward-rightward, backward-leftward, leftward, and forward-leftward was adversely affected (Ku et al., 2016). Furthermore, the ageing effect significantly influenced dynamic stability. Young adults demonstrated better directional stability control in the forward, backward, leftward, and rightward directions compared to the elderly (Faraldo-Garcia et al., 2016). This shown that ageing results in the deterioration of muscle strength and functional ability and impairs the dynamic stance stability.

For the dynamic stance control, the ankle joint is mainly used to regulate the anteroposterior movement, whereas the hip joint is regulating the mediolateral movement (Horak and Nashner, 1986). The ankle joint may limit the directional stability control in the leaning direction of leftward, rightward, and backward (Ku et al., 2016). The upright stance alignment will limit the functional ranges of movement of the knee and the hip joints are activated to compensate for the inadequate ankle joint torque to control lateral balance in dynamic stance. Ankle plantar flexor is responsible to generate plantarflexion and dorsiflexion torques to control the backward and forward trunk leaning, respectively. The range of motion for plantar flexion is ankle movement downward around 0–50° and dorsiflexion is ankle movement upward around 0–20°. Sufficient force and motion by the plantarflexion enable the ankle downward movement to lean the body trunk to the forward direction, to revert to backward leaning. Moreover, the trunk lean direction was associated with ankle muscle weight loading in dynamic stance control (Kanakis et al., 2014). Takahashi et al. (2006) revealed that the forward trunk leaning with and without external weight causes an increased spinal loading. Antagonist co-contraction helps to increase joint stiffness and stabilization in performing motor tasks (Hansen et al., 2002; Geertsen et al., 2013). Nevertheless, the changes in ankle muscle co-contractions among the elderly are affected by postural stability deterioration (Vette et al., 2017).

Motor control is required to regulate the body segment in the dynamic stance control. Rectus femoris is the main quadriceps muscle used to regulate the hip flexion contracture, and its antagonistic muscle is the biceps femoris muscle. Ku et al. (2016) suggest that there is a trend of the increased rectus femoris muscle activity associated with the increased tilt angle of the dynamic platform surface among middle-aged adult (Figure 9.4). Other than the rectus femoris and biceps femoris muscles, the hip abductor muscles group helps to regulate the lateral trunk balance (de Freitas et al., 2009). The gluteus medius muscle controls the lateral pelvic tilt, whereas the gluteus maximus and gluteus minimus muscles control the anterior pelvic tilt. The decrement in ankle muscle strength due to the ageing effect would diminish the muscle activity and stability limits capacity ( Teasdale and Simoneau, 2001; Melzer et al., 2009).

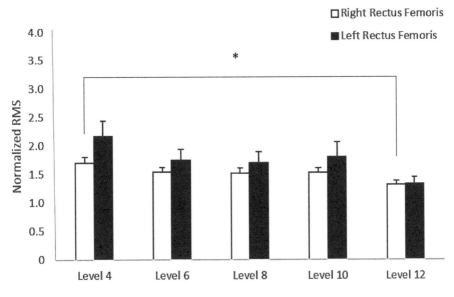

**FIGURE 9.4** The mean RMS of rectus femoris muscle activation in different dynamic stability levels. (Reproduced from Ku, P.X., Abu Osman, N.A., Wan Abas, WAB, 2016. The limits of stability and muscle activity in middle-aged adults during static and dynamic stance. Journal of Biomechanics 49, 3,943–3,948.)

For other muscles involved, the medial gastrocnemius muscle (forward leaning) and biceps femoris muscle (backward leaning) are used in dynamic stance control. The immediate changes of the centre of gravity will stimulate the calf and biceps femoris muscle activation in the dynamic voluntary inclination. The left–right muscle activity asymmetry was found in the medial gastrocnemius and biceps femoris muscles (Ku et al., 2016). Jacobs et al. (2005) suggested that this asymmetry might be due to the total force or strength exerts by the preference leg muscle.

As mentioned in Section 9.3.1, no significant impact of leg dominance on the static stance control. For the dynamic balance control, people with right dominant leg are more likely to shift the weight loading towards forward-rightward, and backward-leftward leaning directions in the LoS test. Since the leg preference is task-dependent, the preference leg for static stance and dynamic stance might not be the same (Huurnink et al., 2014).

### 9.3.3 STATIC STANCE CONTROL WITH TOE EXTENSION

The human foot is comprised mainly of various joints and bones, and all are connected with muscles, tendons, and three layers of ligaments (Wright et al., 2012). The foot is used as a propulsive lever and shock absorber in postural control (Chou et al., 2009). In the foot mechanism, normal foot arch is used to absorb shock, weight-bearing and propulsion phase of gait, while the toes act to maintain body equilibrium by relying on the proprioceptive inputs. A person experienced 60% of weight-bearing pressure distribution at the heel, 8% at the mid-foot, and 28% at the forefoot

(Cavanagh et al., 1987). Many studies have investigated the balance control mechanism in different toe-related quiet stance postures: heel-toe standing, half-toe standing, full-toe standing, and toe extension standing (Nolan and Kerrigan, 2004; Ku et al., 2016; Wan et al., 2019). Amongst all standing posture, toe extension standing is a voluntary action that lifts the toes away from the floor.

The toe extension may result in foot arch elevation, lateral rotation of lower limb, tightness of plantar aponeurosis, or supination at the posterior side of the foot (Hicks, 1954). According to the windlass mechanism, the maximum extension of the toes leads to the tightening of the plantar aponeurosis, further elevates the medial longitudinal arch, and lifts the metatarsal head. The motion in the metatarsophalangeal joints increases the foot pressure against the ground surface. Hicks (1954) indicated that the extended toe induces the foot pressure at the metatarsophalangeal joints, further diminish the postural control. However, this is inconsistent with other studies which reported that no significant evidence was found to prove the effect of toe extension on postural control (Ku et al., 2012b; Yamauchi and Koyama, 2019). Ku et al. (2012b) reported that there is a slightly higher but not significant change of postural sway for toe-extension stance. This may be due to higher effort of balance recovery in a smaller BoS for toe-extension stance. Besides, Yamauchi and Koyama (2019) stated that even the toe flexor did not affect the postural control among young adults, it could be passively used to strengthen the foot arch structure for weight-bearing.

A significant increase of mediolateral sway in the toe-extension stance compared to bipedal stance, and no changes were found for the anteroposterior sway. It is consistent with previous studies that the mediolateral sway is normally small than anteroposterior sway in balance recovery (Pereira et al., 2008; Ku et al., 2012a). It seems like the anteroposterior sway remains even if the toes are extended during the toe-extension stance. Hence, the mediolateral postural control should be considered an important balance component for stance-related studies.

Moreover, some studies claimed that normal adults showed no gender difference on postural stability in bipedal stance, unipedal stance, and toe-extension stance postures (Hageman et al., 1995; Ku et al., 2012b). However, the gender effect may vary amongst different target populations. Sell et al. (2018) found that female soldiers displayed greater static postural control compared to male soldiers, but no differences in dynamic postural control. Mickle et al. (2010) claimed that boys exert greater postural sway compared to girls with similar body weight in both bipedal stance and unipedal stance postures. The greater postural sway in boys may be due to the intrinsic differences of anthropometric characteristics (Chiari et al., 2002). The Fractional Brownian motion modelling explained that males are more likely to experience a higher moment of inertia, and the natural body frequency is higher for females (Farenc et al., 2003; Lee and Lin, 2007). Therefore, females exhibited better postural control than males according to the motion model.

Different toe-related stance postures may require different balance approaches to maintain the body equilibrium. The investigation of toe-extension stance could be used as the active toe-extension stance to compare with passive toe extension that is caused by burn injuries. It would provide normative stability data for people who are concerned with the effect of toe extension, in order to gain a better understanding and improve the quality of life for humans.

## 9.4   CONCLUSION

This chapter provides an overview of postural stance control. Postural stance is the basis of locomotion to perform various tasks. It involves various complex system sustain the body equilibrium against balance disturbance. Many factors contribute to the postural control mechanism, including sensory strategies, movement strategies, biomechanical task constraint, controls of dynamic, orientation in space, cognitive processing and other factors. The postural balance for static stance, dynamic stance and toe-extension stance was discussed. Postural control in static stance is most commonly applied in daily activities, compared to dynamic stance and toe-extension stance postures. Different balance factors may be associated with postural stance control in different stance postures. Therefore, this chapter may help to provide a better understanding of postural control and stance-phase stability.

## REFERENCES

Alexandrov, A., Frolov, A., Horak, F.B., Carlson-Kuhta, P., Park, S., 2005. Feedback equilibrium control during human standing. *Biological Cybernetics* 93, 309–322.

Aniansson, A., Hedberg, M., Henning, 1986. Muscle morphology, enzymatic activity and muscle strength in elderly men: a follow up study. *Muscle Nerve* 9, 585–591.

Aurichio, T.R., Rebelatto, J.R., Castro, A.P.D., 2011. The relationship between the body mass index (BMI) and foot posture in elderly people. *Archives of Gerontology and Geriatrics* 52, e89–e92.

Barrett, R.S., Cronin, N.J., Lichtwark, G.A., Mills, P.M., Carty, C.P., 2012. Adaptive recovery responses to repeated forward loss of balance in older adults. *Journal of Biomechanics* 45, 183–187.

Bisdorff, A.R., Wolsley, C.J., Anastasopoulos, D., Bronstein, A.M., Gresty, M.A., 1996. The perception of body verticality (subjective postural vertical) in peripheral and central vestibular disorders. *Brain* 119(Pt 5), 1,523–1,534.

Blaszczyk, J.W., Cieslinska-Swider, J., Plewa, M., Zahorska-Markiewicz, B., Markiewicz, A., 2009. Effects of excessive body weight on postural control. *Journal of Biomechanics* 42, 1,295–1,300.

Blaszczyk, J.W., Orawiec, R., Duda-Klodowska, D., Opala, G., 2007. Assessment of postural instability in patients with Parkinson's disease. *Experimental Brain Research* 183, 107–114.

Bougard, C., Davenne, D., 2014. Morning/evening differences in somatosensory inputs for postural control. *BioMed Research International* 2014, 287436.

Butler, M., Norton, R., Lee-Joe, T., Cheng, A., Campbell, A.J., 1996. The risks of hip fracture in older people from private homes and institutions. *Age and Ageing* 25, 381–385.

Cavanagh, P.R., Rodgers, M.M., Liboshi, A., 1987. Pressure distribution under symptom-free feet during barefoot standing. *Foot Ankle* 7, 262–276.

Chiari, L., Rocchi, L., Cappello, A., 2002. Stabilometric parameters are affected by anthropometry and foot placement. *Clinical Biomechanics* 17, 666–677.

Chou, S.W., Cheng, H.Y., Chen, J.H., Ju, Y.Y., Lin, Y.C., Wong, M.K., 2009. The role of the great toe in balance performance. *Journal of Orthopaedic Research* 27, 549–554.

Choy, N.L., Brauer, S., Nitz, J., 2003. Changes in postural stability in women aged 20 to 80 years. *The Journals of Gerontology Series A: Biological Sciences and Medical Sciences* 58A, 525–530.

Clark, K.N., 2004. Balance and strength training for obese individuals. *ACSM'S Health Fitness Journal* 8, 14–20.

Collins, J.J., Luca, C.J., 1993. Open-loop and closed-loop control of posture: a random-walk analysis of center-of-pressure trajectories. *Experimental Brain Research* 95, 308–318.

Colne, P., Frelut, M.L., Peres, G., Thoumie, P., 2008. Postural control in obese adolescents assessed by limits of stability and gait initiation. *Gait & Posture* 28, 164–169.

Compston, J.E., Flahive, J., Hosmer, D.W., Watts, N.B., Siris, E.S., Silverman, S., Saag, K.G., Roux, C., Rossini, M., Pfeilschifter, J., Nieves, J.W., Netelenbos, J.C., March, L., LaCroix, A.Z., Hooven, F.H., Greenspan, S.L., Gehlbach, S.H., Diez-Perez, A., Cooper, C., Chapurlat, R.D., Boonen, S., Anderson, F.A., Jr., Adami, S., Adachi, J.D., 2014. Relationship of weight, height, and body mass index with fracture risk at different sites in postmenopausal women: the Global Longitudinal study of Osteoporosis in Women (GLOW). *Journal of Bone and Mineral Research* 29, 487–493.

Corbeil, P., Simoneau, M., Rancourt, D., Tremblay, A., Teasdale, N., 2001. Increased risk of falling associated with obesity: mathematical modeling of postural control. *IEEE Transactions on Neural Systems and Rehabilitation Engineering* 9, 126–136.

Cruz-Gómez, N.S., Plascencia, G., Villanueva-Padrón, L.A., Jáuregui-Renaud, K., 2011. Influence of obesity and gender on the postural stability during upright standing. *Obesity Facts* 4, 212–217.

de Freitas, P.B., Freitas, S.M., Duarte, M., Latash, M.L., Zatsiorsky, V.M., 2009. Effects of joint immobilization on standing balance. *Human Movement Science* 28, 515–528.

de Oliveira, C.B., de Medeiros, I.R., Frota, N.A., Greters, M.E., Conforto, A.B., 2008. Balance control in hemiparetic stroke patients: main tools for evaluation. *Journal of Rehabilitation Research and Development* 45, 1,215–1,226.

Del Porto, H., Pechak, C., Smith, D., Reed-Jones, R., 2012. Biomechanical effects of obesity on balance. *International Journal of Exercise Science* 5, 1.

Di Giulio, I., & Baltzopoulos, V. (2019). Attainment of quiet standing in humans: Are the lower limb joints controlled relative to a misaligned postural reference? *Frontiers in Physiology*, 10, 625.

Drake, R., Vogl, A.W., Mitchell, A.W.M., 2009. *Gray's anatomy for students.* Churchill Livingstone/Elsevier, Philadelphia, PA.

Faraldo-Garcia, A., Santos-Perez, S., Crujeiras, R., Soto-Varela, A., 2016. Postural changes associated with ageing on the sensory organization test and the limits of stability in healthy subjects. *Auris Nasus Larynx* 43, 149–154.

Farenc, I., Rougier, P., Berger, L., 2003. The influence of gender and body characteristics on upright stance. *Annals of Human Biology* 30, 279–294.

Geertsen, S.S., Kjær, M., Pedersen, K.K., Petersen, T.H., Perez, M.A., Nielsen, J.B., 2013. Central common drive to antagonistic ankle muscles in relation to short-term co-contraction training in nondancers and professional ballet dancers. *Journal of Applied Physiology* 115, 1,075–1,081.

Gill, J., Allum, J.H., Carpenter, M.G., Held-Ziolkowska, M., Adkin, A.L., Honegger, F., Pierchala, K., 2001. Trunk sway measures of postural stability during clinical balance tests: effects of age. *The Journals of Gerontology Series A: Biological Sciences and Medical Sciences* 56, M438–447.

Goel, N., Basner, M., Rao, H., Dinges, D.F., 2013. Circadian rhythms, sleep deprivation and human performance. *Progress in Molecular Biology and Translational Science* 119, 155–190.

Greve, J., Alonso, A., Bordini, A.C.P.G., Camanho, G.L., 2007. Correlation between body mass index and postural balance. *Clinics* 62, 717–720.

Hafstrom, A., Fransson, P.A., Karlberg, M., Ledin, T., Magnusson, M., 2002. Visual influence on postural control, with and without visual motion feedback. *Acta Oto-Laryngologica* 122, 392–397.

Hageman, P.A., Leibowitz, J.M., Blanke, D., 1995. Age and gender effects on postural control measures. *Archives of Physical Medicine and Rehabilitation* 76, 961–965.

Hall, S., Myers, M.A., Sadek, A.-R., Baxter, M., Griffith, C., Dare, C., Shenouda, E., Nader-Sepahi, A., 2019. Spinal fractures incurred by a fall from standing height. *Clinical Neurology and Neurosurgery* 177, 106–113.

Han, J., Anson, J., Waddington, G., Adams, R., Liu, Y., 2015. The role of ankle proprioception for balance control in relation to sports performance and injury. *BioMed Research International* 2015, 842804.

Hansen, S., Hansen, N.L., Christensen, L.O., Petersen, N.T., Nielsen, J.B., 2002. Coupling of antagonistic ankle muscles during co-contraction in humans. *Experimental Brain Research* 146, 282–292.

Hernández, A., Slider, A., Heiderscheit, B.C., Thelen, D.G., 2009. Effect of age on center of mass motion during human walking. *Gait & Posture* 30, 217–222.

Hicks, J.H., 1954. The mechanics of the foot II: the planter aponeurosis and the arch. *Journal of Anatomy* 88, 25–30.

Himes, C.L., 2000. Obesity, disease and functional limitation in later life. *Demography* 37, 73–82.

Horak, F.B., 2006. Postural orientation and equilibrium: what do we need to know about neural control of balance to prevent falls? *Age and Ageing* 35, ii7–ii11.

Horak, F.B., Dimitrova, D., Nutt, J.G., 2005. Direction-specific postural instability in subjects with Parkinson's disease. *Experimental Neurology* 193, 504–521.

Horak, F.B., Henry, S.M., Shumway-Cook, A., 1997. Postural perturbations: new insights for treatment of balance disorders. *Physical Therapy* 77, 517–533.

Horak, F.B., Nashner, L.M., 1986. Central programming of postural movements: adaptation to altered support-surface configurations. *Journal of Neurophysiology* 55, 1,369–1,381.

Hue, O., Simoneau, M., Marcotte, J., Berrigan, F., Doré, J., Marceau, P., Marceau, S., Tremblay, A., Teasdale, N., 2007. Body weight is a strong predictor of postural stability. *Gait & Posture* 26, 32–38.

Huurnink, A., Fransz, D.P., Kingma, I., Hupperets, M.D., van Dieen, J.H., 2014. The effect of leg preference on postural stability in healthy athletes. *Journal of Biomechanics* 47, 308–312.

Hytonen, M., Pyykko, I., Aalto, H., Starck, J., 1993. Postural control and age. *Acta Oto-Laryngologica* 113, 119–122.

Jacobs, C., Uhl, T.L., Seeley, M., Sterling, W., Goodrich, L., 2005. Strength and fatigability of the dominant and non-dominant hip abductors. *Journal of Athletic Training* 40, 203–206.

Jáuregui-Renaud, K. (2013). *Postural balance and peripheral neuropathy. Peripheral neuropathy-a new insight into the mechanism, evaluation and management of a complex disorder.* Rijeka: InTech, 125–146.

Jorgensen, M.G., Rathleff, M.S., Laessoe, U., Caserotti, P., Nielsen, O.B., Aagaard, P., 2012. Time-of-day influences postural balance in older adults. *Gait & Posture* 35, 653–657.

Josson, E., Seiger, A., Hirschfeld, H., 2004. One-leg stance in healthy young and elderly adult: a measure of postural steadiness. *Clinical Biomechanics* 19, 688–694.

Kanakis, I., Hatzitaki, V., Patikas, D., Amiridis, I.G., 2014. Postural leaning direction challenges the manifestation of tendon vibration responses at the ankle joint. *Human Movement Science* 33, 251–262.

Keys, A., Fidanza, F., Karvonen, M.J., Kimura, N., Taylor, H.L., 1972. Indices of relative weight and obesity. *Journal of Chronic Diseases* 25, 329–343.

Kisner, C., Colby, L.A., 2002. *Therapeutic exercise: foundations and techniques.* F. A. Davis Company, Philadelphia, PA.

Klinitzke, G., Steinig, J., Blüher, M., Wagner, B., 2012. Obesity and suicide risk in adults – A systematic review. *Journal of Affective Disorders* 145, 277–284.

Kolarova, B., Janura, M., Svoboda, Z., Elfmark, M., 2013. Limits of stability in persons with transtibial amputation with respect to prosthetic alignment alterations. *Archives of Physical Medicine and Rehabilitation* 94, 2,234–2,240.

Ku, P.X., Abu Osman, N.A., Wan Abas, W.A., 2014. Balance control in lower extremity amputees during quiet standing: a systematic review. *Gait & Posture* 39, 672–682.

Ku, P.X., Abu Osman, N.A., Wan Abas, W.A.B., 2016. The limits of stability and muscle activity in middle-aged adults during static and dynamic stance. *Journal of Biomechanics* 49, 3,943–3,948.

Ku, P.X., Abu Osman, N.A., Yusof, A., Wan Abas, W.A., 2012a. Biomechanical evaluation of the relationship between postural control and body mass index. *Journal of Biomechanics* 45, 1,638–1,642.

Ku, P.X., Abu Osman, N.A., Yusof, A., Wan Abas, W.A., 2012b. The effect on human balance of standing with toe extension. *PLoS One* 7, e41539.

Kuo, A.D., Zajac, F.E., 1993. Human standing posture: multi-joint movement strategies based on biomechanical constraints. *Progress in Brain Research* 97, 349–358.

Lee, A.J., Lin, W.H., 2007. The influence of gender and somatotype on single-leg upright standing postural stability in children. *Journal of Applied Biomechanics* 23, 173–179.

Ludwig, O., Kelm, J., Hammes, A., Schmitt, E., Fröhlich, M., 2020. Neuromuscular performance of balance and posture control in childhood and adolescence. *Heliyon* 6, e04541.

Marchetti, G.F., Whitney, S.L., Redfern, M.S., Furman, J.M., 2011. Factors associated with balance confidence in older adults with health conditions affecting the balance and vestibular system. *Archives of Physical Medicine and Rehabilitation* 92, 1,884–1,891.

McClenaghan, B.A., Williams, H.G., Dickerson, J., Dowda, M., Thombs, L., Eleazer, P., 1996. Spectral characteristics of ageing postural control. *Gait & Posture* 4, 112–121.

McGraw, B., McClenaghan, B.A., Williams, H.G., Dickerson, J., Ward, D.S., 2000. Gait and postural stability in obese and non-obese prepubertal boys. *Archives of Physical Medicine and Rehabilitation* 81, 484–489.

Medina, J.J., 1996. *The clock of ages*. Cambridge University, New York.

Mehdikhani, M., Khalaj, N., Chung, T.Y., Mazlan, M., 2014. The effect of feet position on standing balance in patients with diabetes. *Proceedings of the Institution of Mechanical Engineers. Part H, Journal of Engineering in Medicine* 228, 819–823.

Melzer, I., Benjuya, N., Kaplanski, J., Alexander, N., 2009. Association between ankle muscle strength and limit of stability in older adults. *Age and Ageing* 38, 119–123.

Menegoni, F., Galli, M., Tacchini, E., Vismara, L., Cavigioli, M., Capodaglio, P., 2009. Gender-specific effect of obesity on balance. *Obesity* 17, 1951–1956.

Mickle, K.J., Munro, B.J., Steele, J.R., 2010. Gender and age affect balance performance in primary school-aged children. *Journal of Science and Medicine in Sport* 14, 243–248.

Mignardot, J.-B., Olivier, I., Promayon, E., Nougier, V., 2010. Obesity impact on the attentional cost for controlling posture. *PLoS One* 5, e14387.

Miller, J.M., 2014. *The encyclopedia of theoretical criminology*. Wiley, Malden, MA.

Moreland, B., Kakara, R., Henry, A., 2020. Trends in non-fatal falls and fall-related injuries among adults aged ≥65 years – United States, 2012–2018. *Morbidity and Mortality Weekly Report* 69, 875–881.

National Center for Injury Prevention and Control, 2019. *CDC WISQARS™ — Your source for U.S. Injury Statistics, October 25, 2019 ed. Centers for Disease Control and Prevention*, National Center for Injury Prevention and Control.

Netter, F.H., 2010. *Atlas of human anatomy*. Sauders Elsevier, Philadelphia, PA.

Nolan, L., Kerrigan, D.C., 2004. Postural control: toe-standing versus heel-toe standing. *Gait & Posture* 19, 11–15.

Park, S., Horak, F.B., Kuo, A.D., 2004. Postural feedback responses scale with biomechanical constraints in human standing. *Experimental Brain Research* 154, 417–427.

Pereira, H.M., Campos, T.F.D., Santos, M.B., Cardoso, J.R., Garcia, M.d.C., Cohen, M., 2008. Influence of knee position on the postural stability index registered by the Biodex Stability System. *Gait & Posture* 28, 668–672.

Peterka, R.J., 2002. Sensorimotor integration in human postural control. *Journal of Neurophysiology* 88, 1,097–1,118.

Pugh, S.F., Heitman, R.J., Kovaleski, J.E., Keshock, C.M., & Bradford, S.H. (2011). Effects of augmented visual feedback and stability level on standing balance performance using the Biodex Balance System. *The Sport Journal*, 14(1).

Riemann, B.L., Lephart, S.M., 2002. The sensorimotor system, part ii: the role of proprioception in motor control and functional joint stability. *Journal of Athletic Training* 37, 80–84.

Sabashi, K., Ishida, T., Matsumoto, H., Mikami, K., Chiba, T., Yamanaka, M., Aoki, Y., Tohyama, H., 2021. Dynamic postural control correlates with activities of daily living and quality of life in patients with knee osteoarthritis. *BMC Musculoskeletal Disorders* 22, 287.

Salsabili, H., Bahrpeyma, F., Forogh, B., Rajabali, S., 2011. Dynamic stability training improves standing balance control in neuropathic patients with type 2 diabetes. *Journal of Rehabilitation Research and Development* 48, 775–786.

Seeley, R.R., Stephens, T.D., Tate, P., 2006. *Essentials of anatomy and physiology*. McGraw-Hill Education, Boston, MA.

Sell, T.C., Lovalekar, M.T., Nagai, T., Wirt, M.D., Abt, J.P., Lephart, S.M., 2018. Gender differences in static and dynamic postural stability of soldiers in the army's 101st airborne division (air assault). *Journal of Sport Rehabilitation* 27, 126–131.

Shier, D., Butler, J., Lewis, R., 2008. *Hole's essentials of human anatomy & physiology*. McGraw-Hill, Boston, MA.

Shumway-Cook, A., Woollacott, M.H., 2007. *Motor control: translating research into clinical practice*, 3rd ed. Lippincott Williams & Wilkins, Philadelphia, PA.

Takahashi, I., Kikuchi, S., Sato, K., Sato, N., 2006. Mechanical load of the lumbar spine during forward bending motion of the trunk-a biomechanical study. *Spine* 31, 18–23.

Teasdale, N., Hue, O., Marcotte, J., Berrigan, F., Simoneau, M., Doré, J., P. Marceau, Marceau, S., Tremblay, A., 2007. Reducing weight increases postural stability in obese and morbid obese men. *International Journal of Obesity* 31, 153–160.

Teasdale, N., Simoneau, M., 2001. Attentional demands for postural control: the effects of ageing and sensory reintegration. *Gait & Posture* 14, 203–210.

Tinetti, M.E., Speechley, M., Ginter, S.F., 1988. Risk factors for falls among elderly persons living in the community. *New England Journal of Medicine* 319, 1,701–1,707.

Tomomitsu, M.S., Alonso, A.C., Morimoto, E., Bobbio, T.G., Greve, J.M., 2013. Static and dynamic postural control in low-vision and normal-vision adults. *Clinics* 68, 517–521.

Vette, A.H., Sayenko, D.G., Jones, M., Abe, M.O., Nakazawa, K., Masani, K., 2017. Ankle muscle co-contractions during quiet standing are associated with decreased postural steadiness in the elderly. *Gait & Posture* 55, 31–36.

Wan, F.K.W., Yick, K.-L., Yu, W.W.M., 2019. Effects of heel height and high-heel experience on foot stability during quiet standing. *Gait & Posture* 68, 252–257.

Winter, 1995. Human balance and posture control during standing and walking. *Gait & Posture* 3, 193–214.

Woollacott, M.H., Tang, P.F., 1997. Balance control during walking in the older adult: research and its implications. *Physical Therapy* 77, 646–660.

World Health Organization, 2008. *WHO global report on falls prevention in older age*. World Health Organization, France.

Wright, W.G., Ivanenko, Y.P., Gurfinkel, V.S., 2012. Foot anatomy specialization for postural sensation and control. *Journal of Neurophysiology* 107, 1,513–1,521.

Yamauchi, J., & Koyama, K. (2019). Toe flexor strength is not related to postural stability during static upright standing in healthy young individuals. *Gait & Posture*, *73*, 323–327.

Yoon, J.J., Yoon, T.S., Shin, B.M., Na, E.H., 2012. Factors affecting test results and standardized method in quiet standing balance evaluation. *Annals of Rehabilitation Medicine*, 36, 112–118.

# 10 Prosthetic Hand using BCI System

*N A Abu Osman*

University of Malaya, Kuala Lumpur, Malaysia

*S Yahud*

Universiti Malaysia Perlis (UniMAP), Kampus Pauh Putra,
Arau, Malaysia

## CONTENTS

DOI: 10.1201/9781003196730-10

## 10.1   INTRODUCTION

A BCI is one of the modalities used to connect artificial organs and natural systems. This communication system allows messages or commands to be conveyed from an individual to external devices without passing through the brain's normal output pathways of peripheral nerves and muscles. A BCI user is trained, and the system learned to recognize the intentional movement based on the neural signals specific to the intention and translate it to the output desired by the user. Hence, BCI application provides a solution for people with severe motor disabilities, and in a worst case scenario, for a completely paralyzed or "locked-in" patient who is unable to communicate in any way. Applications of BCI are wide and could be varied depending on the needs of the user, such as communication, controlling the environment, moving a prosthetic limb, or rehabilitation. In this study, the BCI system is used to control a prosthetic hand.

## 10.2   PROSTHETIC HAND

Development of prosthetic hand is inspired by the functionality and aesthetic of the human hands. The human hand is a very complex and malleable tool connected to the most powerful and complicated controller, the brain. The hand is made up of the wrist, palm and fingers. The first finger is the thumb followed by the index, middle, ring and little finger. The skeleton of the hand can be divided into three segments: the carpus or bones of the wrist, the metacarpus or bones of the palm, and the phalanges or bones of the fingers.

Hand prostheses can be classified into five different types: passive, hybrid, task specified, body powered and externally powered. To determine the type of prosthesis suitable for each amputee, one must consider the level of amputation and understand the purpose of having a prosthetic hand. The passive prosthetic hand is the most popular choice for cosmetic restoration, but it does not provide any functional task. In this study, below-elbow amputation is the chosen level, thus the hybrid prosthetic hand is not an option because it is usually built for above-elbow amputation. Most early versions of functional below-elbow prostheses known are body-powered and relied on the contralateral hand or relative motion between the shoulder, upper arm and forearm for the operation. Body powered below-elbow prostheses usually consist of split hook or five fingers, socket, and harness. This type of prosthetic hand only provides open and close hand action. Studies were carried out to overcome these limitations by switching to the externally powered prosthetic hand. A task specified prosthetic hand also gives limited action and has restricted the performance.

Several externally powered prosthetic hands exist, ranging from DC motor, ultrasonic motor, pneumatic cylinder, hydraulic cylinder, shape memory alloy material (SMA) and electroactive polymer (EAP). Advances in robotic technology have been helpful in improving the control and design of prosthetic hands. There is a close relationship between robotics and prosthetic hands since both provide human like motion and prehension. In a robotic hand, the focus is to imitate the function of the human hand and enhance the performance in a way appreciated by the human. Thus, in robotic application two fingers are sufficient. However, three fingers are needed to perform dexterous tasks in an unstructured environment and to achieve full grasp of two-dimensional objects.

## 10.3 LITERATURE REVIEW

### 10.3.1 ANATOMY OF THE HUMAN HAND: INTRODUCTION

The human hand is made up of the wrist, palm and fingers. First finger is the thumb, followed by the index, middle, ring and little fingers. The skeleton of the hand can be divided into three segments: the carpus (or bones of the wrist), the metacarpus (bones of the palm), and the phalanges (bones of the finger), as shown in the Figure 10.1.

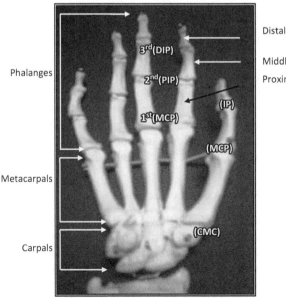

**FIGURE 10.1** Anatomy of the human hand (photographed by the author).

There are four bones that make up each finger (except the thumb): the metacarpal, proximal phalanx, middle phalanx and distal phalanx. These are connected at three joints. The first joint connects the proximal phalanx to the metacarpal of each respective finger. It is called the metacarpophalangeal (MCP) joint. The second joint is the proximal interphalangeal (PIP) joint, which connects the proximal phalanx to the middle phalanx. The third joint is distal interphalangeal (DIP) joint, which connects the middle to distal phalanx. The thumb is made up of three bones: the first metacarpal, proximal phalanx and distal phalanx. The first metacarpal is connected to the bone of the wrist at the carpometacarpal (CMC) joint. The proximal phalanx of the thumb is connected to the first metacarpal at the MP joint and the distal phalanx is connected to the proximal phalanx at the interphalangeal (IP) joint.

All the interphalangeal (IP) joints are hinge joints. A hinge joint allows movement in one plane: in this case, flexion and extension. The metacarpal (MP) joint is more complex than the IP joint and permits additional movements: abduction, adduction and circumduction. The abduction and adduction or side-to-side movements are very limited and cannot be performed when the finger is flexed. Circumduction is the movement where the bone is made to circumscribe a conical space. The carpometacarpal (CMC) joint of the thumb permits flexion, extension, abduction, adduction, circumduction and opposition. Opposition is the movement when the tip of the thumb is brought to meet the volar surface of every flexed finger.

The member of degree of freedoms (DOFs) describes the flexibility of motion of a joint. In artificial manipulator, DOFs is a terminology used to describe the number of directions that a robot can pivot or move a joint. An ideal artificial hand possesses at least four DOFs in each finger and five DOFs in the thumb, therefore resulting in a total of 21 DOFs in the hand, not considering the movements of the wrist and movements of the fourth and fifth metacarpal bones. All the IP joints are one DOF joint, whereas the MP and CMC joints are both two DOFs joint. Functions of the human hand are realized by 17 articulations and 19 muscles situated entirely within the hand, and about the same number of tendons activated by the forearm's muscles. Human grasps are divided into two types: power and precision. The power grasp is applied when requiring power for handling an object, and the power is increased as the size of the object increases. Examples of power grasps are circular and cylindrical. The precision grasp is for fine handling, with emphasis on increase in dexterity. The acts of precision grasp include tripod pinch, key pinch and tip pinch. With the knowledge of the anatomy of the human hand and an understanding of the range of motion, researchers will have a better idea on how the artificial hand can be developed to be close to the human hand, both structurally and functionally.

### 10.3.2 Prosthetic Hand: An Introduction

Losing a hand is tragic for any human being. An individual without a limb is known as an amputee. There are two types of hand amputation: below-elbow and above-elbow. The major challenge faced by every below-elbow amputee is the difficulty to perform several activities of daily living (ADL) such as dressing, feeding and toileting. Amputation occurs due to many reasons such as mishap during war or motor accidents, diseases and congenital anomalies. Studies in prosthetics have been carried out to find an artificial substitute for the missing hand, known as a prosthetic hand.

The history of mankind has shown that man's attempts to replace functions of the hand and arm extends at least as far back as the time of the second Punic War (218– 201 BC) when Marcus Sergius, who lost his right hand, was provided with an iron one. However, recognition of the work was only given after World War I, in the USA and Great Britain, where it was a general rule to supply each eligible beneficiary with a work arm and a dress arm. Since then, research efforts have been carried out by many universities and organizations in improving the artificial hand. Artificial hands that functioned as prosthetic hand can be classified into passive, body powered, externally powered, hybrid and task specific prostheses.

A passive prosthetic hand is basically worn for cosmetic restoration. This type of prosthetic hand is designed to look like a human hand and often fabricated based on the missing limb. It gives only aesthetic appearance needed by the amputee, but it does not help in performing tasks that are often executed by the human hand. Another type of the prosthetic hand is the hybrid prosthesis. A hybrid prosthesis is generally fabricated for those with above-elbow amputations. It comprises both body-powered and electrically-powered components at either the device terminal or the arm. Some activities require a purpose-built prosthesis due to limitations in the performance or the susceptibility to damage of other types of prosthesis. Therefore, task-specified prostheses are often built for recreational purposes, considering activities as diverse as music, water skiing, cycling, fishing, weightlifting and gymnastics.

Most of the ancient below-elbow prosthetic hands known are body powered and relied on the contralateral hand or relative motion between the shoulder, upper arm and forearm for the operation. A body powered below-elbow prosthesis usually con- sists of a split hook or five fingers, a socket and a harness. Beside the design which is not anthropomorphic, the socket that fits the stump and harness that fits across the shoulder causes skin irritation and an uncomfortable feeling. One example of an advanced body powered prosthetic hand is from the WILMER group. It was designed to eliminate harness, easy for donning and doffing, and with socket adjustability to compensate growth. It is especially designed for children. Opening or closing of the body powered terminal device is controlled by movements of the elbow: opening is coupled to the elbow extension and closing is coupled to elbow flexion. However, this type of prosthetic hand only provides open and close hand actions, and due to this limitation studies are carried out to overcome it by switching to the externally powered prosthetic hands.

The current trend in prosthetic hands is the externally powered prosthesis. External actuators such as DC motors, ultrasonic motors, pneumatic cylinder, hydraulic cylin- der, and shape memory alloy material (SMA) are used to operate the hand. The moti- vation of a prosthetic hand is to recover the functions of the human hand, and therefore it should be able to be worn and carried around by the user. The success of any prosthetic limb is depending on acceptance by the user. To be accepted by the user, a prosthetic hand should possess the look and weight like the human hand as well as function like a human hand. Besides, it must be easy to operate and user-friendly. Several methods have been used to preserve the communication between a user and the prosthetic hand. The most common method is to take the electromyography (EMG) signals acquired from the remaining limb and input it into the controller of the hand. However, research studies have also conducted to explore the possibilities of using mechanomyography (MMG) signals or electroencephalogram (EEG) signals to operate the hand.

### 10.3.3 Review on Artificial Hands

Research on artificial hands is not restricted to prosthetic usage, and today it has emerged into a variety of applications. Therefore, multi-fingered artificial hands have been developed in many countries, including the US, the UK, Germany, France, China and Japan. There are many reasons why we need to develop an artificial hand. Wide applications of artificial hands have catalyzed its development. Artificial hand has been used in the industry to execute manufacturing functions at large scale, low cost and with great accuracy. In research applications, artificial hands are often used to replace the human workers to perform tasks in a hazardous environment. Artificial hands built specifically for space applications are designed with tools and control interfaces meant to be manipulated by human hands. Artificial hands designed for rehabilitation purposes are used to restore the function of the human hands for amputees and locked-in patients.

The development of the artificial hand is inspired by the functionality and dexterity of the human hand. Furthermore, technology advancement has driven the artificial hand to evolve technologically and becoming more like a robotic hand. The evolving pattern in the development of artificial hands showed the increasing number in the DOF and increasing number of functional artificial fingers per hand. Actuators used to actuate the hand varied, depending on the need and purpose of the hand. DC motors are commonly used as actuators, however other type of actuators is also available, such as SMA, and pneumatic. Study and development of the prototype of artificial hands (referring to prosthetic and robotic hands) shows an increasing trend in the past ten years. Most of the patents are from North America and few are from European countries, Japan, China and Australia.

### 10.3.4 Review on the Controller

The central nervous system (CNS) controls human movement by sending signals to each muscle via $\alpha$-motor neurons. This signal then activates the muscle and yields tensions in it, resulting in joint torque. The human body will then accomplish the required motion. The human movement can be classified into three levels: muscle level, joint level and motion level. Amputees are not able to perform a motion due to the absent parts or tools, and in this case, the hand. The primary concern in the application of the prosthetic hand is how to communicate with it and to control it effectively. Amputees have several options that enable them to communicate and further control the prosthetic hand. Those who are healthy and conscious can wear a prosthetic hand and use either electromyography (EMG) or mechanomyography (MMG) signals from the residual limb to control it. However, for those who have severe motor disabilities, or are totally 'locked-in', they are unable to communicate by any means. The brain signals are the only signals available and can be regulated for communication purposes. This method of using electroencephalography (EEG) signals enable movement to be restored for people with severe motor disabilities (e.g., amyotrophic lateral sclerosis, cerebral palsy, spinal cord injury). Subsections 2.3.2–2.3.4 review three existing methods of using EMG, MMG and EEG signals as input for the controller in the prosthetic hands.

## 10.3.5   ELECTROMYOGRAPHY (EMG) SIGNALS

Most of the existing systems used to control the prosthetic hands are based on the EMG signals generated by the residual limb. The EMG signal is a small electrical current resulting from the activation of individual muscle fibres during contractions. Raw EMG signals captured are then amplified, rectified and filtered accordingly to produce control signals. Waseda hand is among the pioneers in the field. It was developed in 1964, utilized EMG as a control signal a year later, and was available commercially in 1978. The MARCUS hand can perform precision and power grips similar to the human hand, with the control signals coming from continuous flexor and extensor EMG signals. Kobe hand estimates torque from the antagonist pair flexor and extensor EMG signals on the forearm and used it as the control signal. Huang and Chen from National Taiwan University utilize EMG signal pattern discrimination to control NTU hand. EMG-controlled prosthetic hand may offer more than one practical DOF (flexion-extension of the forearm muscles), generally not more than two DOFs. This number may be limited by the ability of the user to learn unnatural movements to activate hand motion and the ability of the controller to decode the resulting EMG. An EMG signal can be further classified using a variable learning rate (VLR) based neural network with parametric Autoregressive (AR) and wavelet transform, to perform grasp function: power grasp, cylinder grasps, grip, and fingertip grasp. However, the EMG-controlled prosthetic hand is still unable to perform precise handling.

## 10.3.6   MECHANOMYOGRAPHY (MMG) SIGNALS

Mechanomyography is the measurement of the mechanical vibrations elicited by contracting muscles. A study on the MMG signals as control signals for the free-standing prosthetic hand in a controlled clinical environment has been carried out successfully. Some advantages of MMG signals over EMG signals are that MMG signal is unaffected by skin impedance, MMG signals have been found to be more accurate predictors of fatigue during muscle contraction, sensor placement does not need to be overly precise, and in terms of cost of fabrication, MMG sensors can be produced at a lower cost than EMG sensors. However, unlike an EMG signal a MMG signal is susceptible to interference by the surrounding noise. MMG-controlled prosthetic hand is not yet satisfactory for practical use since it is only able to perform open and close hand motions.

## 10.3.7   ELECTROENCHEPHALOGRAM (EEG) SIGNALS

The wish of being able to use brain signals to communicate with external devices is the main objective in the BCI research. BCI is a communication system in which messages or commands are conveyed from an individual to an external device without passing through a brain's normal output pathways of peripheral nerves and muscles. Brain signals could be acquired using both invasive and non-invasive methods. Non-invasive method is commonly used because it is easy to handle and has a minimal risk of infection. The brain signals or electroencephalography (EEG) acquired are further processed and classified according to the required tasks. The type of EEG signals used for controlling the prosthetic hand in BCI application are oscillatory EEG; event-related desynchronization (ERD) and event-related synchronization (ERS) components because of actual and imagined hand movement.

BCI application is wide and can be varied depending on the need of the user, such as communication, controlling the environment, or moving the prosthetic limb. Birch and Mason (2000) used the LF-ASD to allow users to navigate a maze by making turning decisions at intersections which could be implemented for a wheelchair control. Graz-BCI is used by a tetraplegic patient to control the opening or closing of a hand orthosis. The innovation of using EEG signals as the input control of a prosthetic hand is still considered to be in the preliminary stage, if compared to the dexterity of the human hand. The aim of utilizing the BCI system as an alternative communication channel to people with severe motor impairment or those who are completely paralyzed or "locked-in" is a milestone in the development of prostheses.

This chapter gives a general idea on how the development of artificial hands has been progressing and the types of available technologies that support the advancement of the prosthetic hand. Applications of artificial hands are many. However, in this study, the development of the artificial hand is specified for prosthetic applications. To be precise, the prosthetic hand developed in this study is to be used by people with severe motor impairment, or are totally locked-in.

## 10.4 METHODOLOGY

The primary concern for a prosthetic hand is to restore functions of the human hand as well as maintaining the humanlike appearance. Therefore, before commencing on the design and fabrication, a comprehensive study on the anatomy of the human hand was conducted. The understanding of the human hand helps in designing the structural components of the artificial hand together with the operating mechanism.

The design process is divided into three parts: structural design, design of the working mechanism, and the fabrication and assembly processes. The functionality of the developed prototype prosthetic hand is evaluated with respect to the objective of the study.

### 10.4.1 DESIGN CRITERIA

The objective of the project is to produce a prosthetic hand which can perform the four essential tasks: cylindrical grasp, key pinch, pulp-to-pulp pinch and tripod pinch. For it to perform the four tasks, the hand should possess the maximum number of DOFs and have fingers configured like human fingers. The proposed design of a prosthetic hand is made up of five fingers, including the thumb. The assigned

**TABLE 10.1**

**Assigned Components of the Prosthetic Hand**

| Components of the Human Finger | Assigned Components |
| --- | --- |
| Distal phalanx | L3 |
| Middle phalanx | L2 |
| Proximal phalanx | L1 |
| First metacarpal (thumb) | L0 |

components of the prosthetic finger as compared to the human finger are shown in Table 10.1.

The basic unit in this design is the prosthetic finger. All four fingers had similar components and are arranged accordingly in the following sequence: **L3** connects to **L2** at the distal interphalangeal (DIP) joint and the other end of **L2** is connected to **L1** at the proximal interphalangeal (PIP) joint. The finger is then mounted on the palm at the metacarpophalangeal (MCP) joint. The thumb, however, is different from the other fingers because thumb does not have the **L2**. The arrangement made for thumb is as follow: **L3** connects to **L1** at the interphalangeal (IP) joint whilst the other end of **L1** connects to **L0** at the metacarpophalangeal (MCP) joint. The thumb joins the hand at the carpometacarpal (CMC) joint. The PIP and DIP joints of the human hand could be modelled as hinge joints. On the other hand, the MCP and CMC joints of the human finger are often modelled as universal joints.

The study of the anatomy of human hand was a great help in designing the prosthetic hand. The proposed design yields a total of 16 DOFs, with each finger having three DOFs and the thumb having four DOFs, including an additional circumduction motion at the CMC joint. Each DOF movement was controlled by an individual actuator, thus the hand uses a total of 16 units of actuator. A simple tendon-drive mechanism was introduced in the design. A string was used to transmit torque from the actuator to the distant joint. The string pulled the segment as the actuator rotated. The angle and speed of the flexion were dependent on the output of the actuator. Flexion of a particular segment was achieved when the string was pulled by the actuator and reset to its original position when the string was released. The reset mechanism was achieved with the use of a torsion spring in the joint. The stored resistive forces inside the torsion spring reset the segment to its original position. The string was functioning as a flexor digitorum profundus tendon of the human hand. The prosthetic hand was equipped with potentiometers as angular position sensors as well as force sensors for tactile sensing. Sensors were included in the design consideration for the control purposes.

## 10.4.2 The Structural Design

The design of an individual component of the finger is inspired by the shape of the human phalanx itself. This is for the functional purposes and to give the prosthetic hand a humanlike appearance. The structural consists of three components: the aluminium plates, the brass spacers and the back cover.

### 10.4.3 The Aluminium Plates

The main component of the prosthetic finger is made from aluminium plates. An aluminium plate is used to model the phalanx for each segment. Each segment is made up of two parallel identical aluminium plate of 1.5mm thickness. The shape of each segment is made like the human finger. Figure 10.2 shows every component that represent each phalanx: **L3**, **L2** and **L1** accordingly. The **L3** (distal phalanx) was cut narrow compared to the other two segments. The inclination is needed to provide an area for the tactile sensor and silicone tip mounting.

The parallel plate's structure allows the finger structure to be easily assembled and disassembled. Furthermore, the design results in a better accuracy and simpler fabrication and machining processes. Figures 10.3 and 10.4 shows the arrangement of the plates to form a prosthetic finger and thumb, respectively.

The length of each segment was determined by the distance between two centres of rotation of the joints. The lengths of **L1**, **L2** and **L3** were respectively determined by the distance between the MCP and PIP joints, PIP and DIP joints, and DIP to the tips of respective finger. The distance between the MCP to PIP (MCP–PIP), PIP to DIP (PIP-DIP) and DIP to tip of the finger (DIP-tip) were denoted by $l_1$, $l_2$ and $l_3$ respectively. The thumb, as described earlier in this chapter, comprises three bones: distal phalanx, proximal phalanx and first metacarpal, and were denoted by **L2**, **L1** and **L0,** respectively. The length of the distal segment, $l_2$ was measured from the centre of the IP joint to the tip of the thumb (IP-tip) and the length of the proximal segment $l_1$ covered the distance between the centre of MCP joint to the IP joint (MCP–IP). The length of the base segment (the metacarpal), $l_0$ was measured from the centre of the MCP joint to the CMC joint (CMC–MCP).

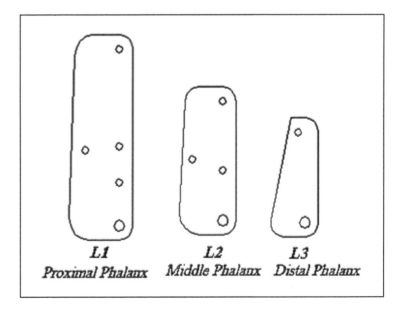

**FIGURE 10.2**  Different sizes for phalanx.

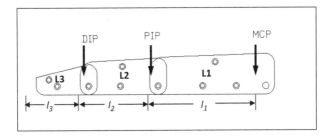

**FIGURE 10.3** Prosthetic finger.

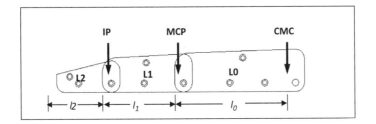

**FIGURE 10.4** The thumb.

Fibonacci sequence applied to the human hand as stated by Littler and quoted by Park et al. (2003) is as follows: the difference between the length of the following phalanx and the sum length of the two prior phalanges must equal to zero. The sum of the distal phalanx and the middle phalanx should be equal to the length of the proximal phalanx. However, in his work, Park has shown that the bone lengths of the finger do not follow the Fibonacci relationship, but the motion paths of the digits still form an equiangular spiral path. Table 10.2 shows the lengths $l_1$ $l_2$, and $l_3$ for every finger, and the ratio for the segments are tabulated in Table 10.2.

## TABLE 10.2
### $l_0$, $l_1$, $l_2$ and $l_3$ Values for All Fingers

| | CMC–MCP – $l_0$ (mm) | MCP–IP – $l_1$ (mm) | IP–tip – $l_2$ (mm) |
|---|---|---|---|
| **Thumb** | 54 | 31 | 27 |
| | MCP–PIP – $l_1$ (mm) | PIP–DIP– $l_2$ (mm) | DIP–tip – $l_3$ (mm) |
| **Index** | 53 | 33 | 24 |
| **Middle** | 57 | 35 | 27 |
| **Ring** | 55 | 33 | 25 |
| **Little** | 42 | 25 | 23 |

Length for the **L1**, **L2** and **L3** are determined from each respective value of $l_1$, $l_2$ and $l_3$ from previous paragraph. The equations to get the values of **L1**, **L2**, and **L3** are as follows:

**L0** = $l_0$ + 8mm (thumb)
**L1** = $l_1$ + 8mm (all fingers including the thumb)
**L2** = $l_2$ + 4mm (thumb)
**L2** = $l_2$ + 8mm (all fingers excluding thumb)
**L3** = $l_3$ + 4mm (all fingers excluding thumb)

### 10.4.4 SPACERS

Spacers were introduced in the design for two reasons: (i) as an anchor to attach two segments together as well as providing sufficient area for the torsion spring insertion, and (ii) as part of the framework of the design to hold the parallel plates together as well as determining the size of the segments. The function of the spacer in the design is shown in Figure 10.5.

### 10.4.5 SPACERS: AN ANCHOR

The hinge joints at the DIP, PIP and MCP were established by insertion of the brass rod at those joint, acting as anchor. Spacers were used in the design to determine and hold the intended width of the segments, thus determining the width of the finger.

### 10.4.6 SPACERS: THE FRAMEWORK

The two parallel plates were fixed together by the shaft placed at each joint. However, the shaft alone was not sufficient to provide rigidity and stability to the structure. Spacers were added into the components so that they could hold the intended width and added rigidity and robustness to the structure. Beside the rigidity and stability of

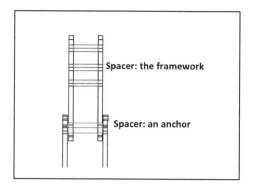

**FIGURE 10.5** Functions of the spacers.

the design, the spacer placed inside the components also provided structure for the tendon to route and transmit the torque produced by an actuator.

### 10.4.7  BACK COVER

The functions of the back cover were to wrap the back part of the component and to act as a stopper to prevent over extension, provided the base for string attachment and to give support to the torsion spring. Figure 10.6 shows the shape of the back cover, but the length of the back cover is dependent on the length of the housing segment. As in the human finger, the degree of extension is rather small compared to flexion. In this design, extension was not taken into consideration since the four focus tasks did not require extension. The back cover was needed for a tendon material attachment, so that when the tendon was pulled by the actuator, the component would follow to flex. As the component flexed, the torsion spring experienced resistive force and built it up within the spring. The back cover helped the spring to store the resistive forces and prevent the return spring mechanism from collapsing.

### 10.4.8  THE WORKING MECHANISM

The working mechanism proposed in the design was inspired by the working principle of the flexor digitorum profundus (FDP) tendon in the DIP joint of the human finger. The FDP tendon pulled the distal phalanx to perform flexes and released it to reset to its original position. In this design, as the component flexed due to the pulling effect from the tendon material, the torsion spring placed at each joint developed resistive force. The stored resistive force within the spring kicked the component to its original position when the tendon released it.

### 10.4.9  TENDON-DRIVE MECHANISM

Each component of the prosthetic finger has its own tendon-drive mechanism and spring return mechanism. Figure 10.7 shows one set of the working mechanism implemented in this design. The terelyne string is pulled by the DC motor and causes the attached component to bend with respect to centre of the joint.

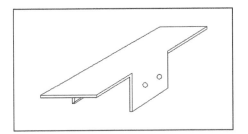

**FIGURE 10.6**   The back cover.

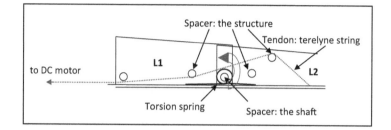

**FIGURE 10.7**   The working mechanism.

### 10.4.10   SPRING RETURN MECHANISM

As the component of the prosthetic finger curled, the torsion spring installed at each joint built up a resistive force within it. The stored resistive force kicks the component back to its original position when the actuator releases the string. The rate of which string is released determines the speed of the component returning to its original position. A helical torsion spring was used for the return mechanism as shown in Figure 10.8.

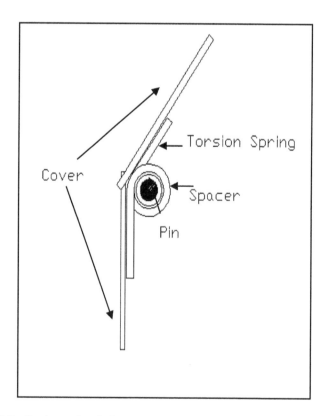

**FIGURE 10.8**   Torsion spring design.

### 10.4.11 ACTUATOR

The proposed design employs a DC micromotor as an actuator for every DOF of motion. DC motors were chosen in this study instead of other actuators because it was easily controlled and required less maintenance when compared to hydraulic or pneumatic actuators.

### 10.4.12 THE ASSEMBLY PROCESS

The whole process of producing a proposed prosthetic hand started with the assembly of an individual finger. Subsequently, the assembled fingers were arranged accordingly onto the 10mm thick aluminium plate, and finally, strings for each component were inserted and connected to the respective DC motors. Potentiometers were installed at every joint of a assembled finger. One finger required three units of potentiometer.

The prosthetic hand was equipped with a force sensor each on the fingertip and volar surface of the middle segment. In this work, force sensors were placed at least at two positions: (i) on the fingertip of the thumb, the index finger, and the middle finger to measure exerted force during pinching, and (ii) on the volar surface of middle phalanx (**L2**) for measuring exerted force during the grasp action.

For the installation of the potentiometer, it was important to identify the moving and static components. All three segments acted as both static and moving components depending on the motion that took placed.

## 10.5  EXPERIMENTS

### 10.5.1  TO TEST THE FUNCTIONALITY AND PERFORMANCE OF THE PROSTHETIC HAND

This experiment was conducted prior to the completion of the prototype. The experiment was carried out in two experiment stages: (i) to test the range of motion attained by the prototype, and (ii) to investigate the ability of the prototype to be used with the BCI system. Two types of sensor were used in the design as feedback control devices. Feedback was obtained from both the potentiometers and force sensors. The potentiometers gave angular displacements of each joint and the FlexiForce® sensor gave the contact force exerted by a finger onto the object handled.

The first stage of the experiment was to investigate a range of motion of the prototype as well as the stability of the movement. The user was required to key-in the desired value of angle for each segment into a graphical user interface (GUI) box displayed on the workstation of the system, as shown in Figure 10.9. The input columns displayed are as follows (from the left):

  (a) little finger
  (b) ring finger
  (c) middle finger
  (d) index finger
  (e) thumb

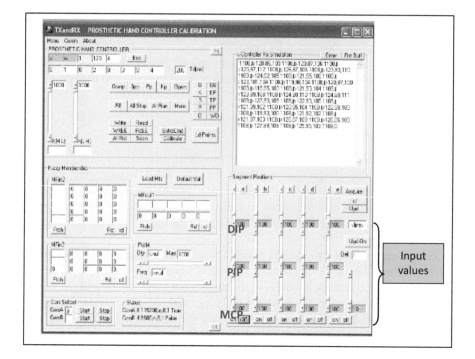

**FIGURE 10.9**  A graphical user interface (GUI) box for fuzzy controller.

The ranges of desired values keyed-in were from 10–90° for each individual segment. This range of values was sufficient for the prototype to perform the four essential tasks. The prototype flexed up to the desired angle or the maximum angle which could be attained if the contact force between object and volar surface did not exceed 453g due to sensor range.

Various tasks were performed to ensure that the hand fulfilled the required objectives: cylindrical grasp, key pinch, pulp-to-pulp pinch and tripod pinch. The experiment setup for this stage of experiment is shown in Figure 10.10.

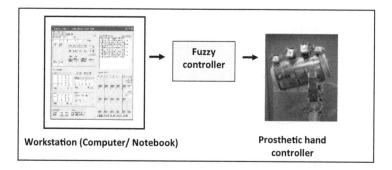

**FIGURE 10.10**   The experimental setup for testing functionality of the prototype.

The second stage of the experiment required the integration between the prototype and the BCI system. A human subject was needed to control the prototype by intend. EEG signals from the subject were acquired using electrodes, as shown in Figure 10.11. Instruction would be given to controller based on the recognized pattern during signal processing. The flow diagram on how the BCI system was used to control the prosthetic hand is shown in Figure 10.12. The prototype was again required to perform the four essential tasks.

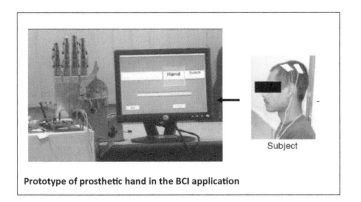

**FIGURE 10.11**    The experimental setup for testing prototype with the BCI system.

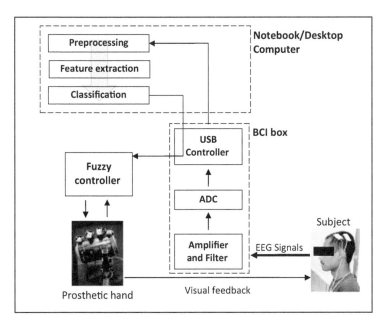

**FIGURE 10.12**    A flow diagram for testing prototype functionality with BCI system.

### 10.5.2  To Investigate the Functional Weight Tolerated by the Prototype

The second experiment was conducted to investigate the functional weight accepted and the robustness of the design. The proof ring method was used to measure the functional weight permissible at each joint during curling. The proof ring was equipped with four units of strain gauges on its inner and outer surfaces of the ring oppositely.

The calibrated proof ring was then used for measuring tension developed in each segment during flexion and rest position. Tendon strings connecting the actuator and the segment were cut in the middle and the proof ring inserted between them. Readings were taken in the rest position, and during flexion at 45° and 90° for each segment.

### 10.5.3  To Evaluate the Motion of the Prosthetic Hand

This experiment was conducted to study the motion path followed by the prosthetic hand. The motion path of an individual finger was captured using video cameras and analyzed using a motion analysis software. A three-dimensional calibration frame was placed in the space in which the motion of interest would occur. The calibration frame served as a global reference system. The calibration frame was then replaced with the prosthetic hand. Reflective markers (5mm in diameter) were pasted onto the prosthetic hand at four locations for every finger. Markers were placed on both side of the fingertip, DIP joint, PIP joint and MCP joint of each finger. Movements made by the prosthetic hand were recorded. Flexion for each finger was recorded, starting with the index and followed by the middle, ring, and finally the little fingers.

## 10.6  RESULTS

As shown in the Figure 10.13, the prosthetic hand is equipped with potentiometers and force sensors. The followings are discussed in this chapter:

  (i)   Functionality of the prosthetic hand
  (ii)  Performance characteristics of the prosthetic hand
  (iii) Trajectory path of the fingertip and joints for each finger

### 10.6.1  Functionality of Prosthetic the Hand (Prosthetic Hand– Controller-BCI Box-Subject)

The first experiment which was done prior completion of the prosthetic hand was the functional test without integration with the BCI system to investigate the range of motion executed by the hand. In this experiment, the prosthetic hand was equipped with a fuzzy controller and feedback sensors. Angular values between 10–90° were set as the working values for each segment.

The potentiometer placed at each joint recorded the angular displacement and fed the value to the controller. Thus, the potentiometer reading at each joint denoted the

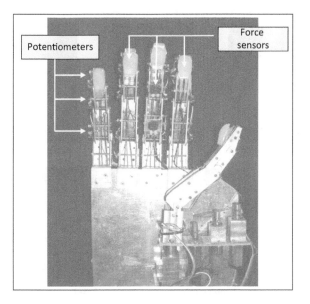

**FIGURE 10.13**   The prototype of the prosthetic hand.

respective flexion angle. The stability of the design was observed throughout the experiment and a sample of the output is displayed on the graph in Figure 10.14. A constant value obtained by each potentiometer at designated angles demonstrates the stability of each joint for a particular segment for the specified angle.

In the second stage of this experiment, the hand was further tested as an external device for the BCI system. The hand related to a fuzzy controller and the proposed

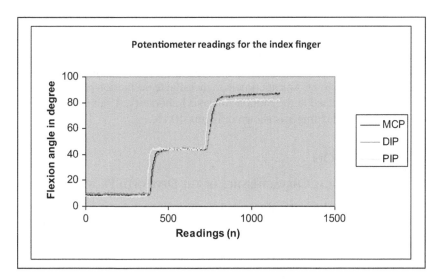

**FIGURE 10.14**   Potentiometer readings for the index finger.

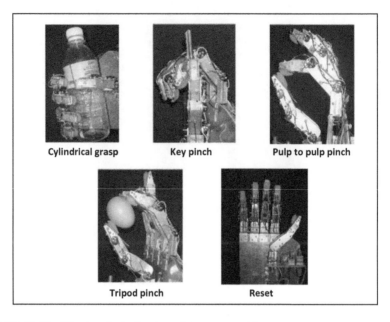

FIGURE 10.15   The four desired tasks and the reset position.

BCI system. A human subject was introduced at this stage. Brain signals acquired from the subject were used to control the prosthetic hand. Results from the experiment are shown in Figure 10.15. The hand could perform the four essential tasks pre-determined in the objective of the study: cylindrical grasps, key pinch, pulp-to-pulp pinch and tripod pinch.

### 10.6.2   TEST TO EVALUATE THE FINGERTIP TRAJECTORY AND MOTION PATH

It is important to evaluate the fingertip trajectory of every prosthetic finger. A theoretical fingertip trajectory was identified earlier in the designing stage. The desired fingertip trajectory of the prosthetic hand was determined by the integral of the motion path for each joint. An experiment to investigate the motion path and fingertip trajectory was conducted in the Motion Analysis Laboratory. A sample of a fingertip trajectories of an index finger is shown in Figure 10.16.

## 10.7   DISCUSSION

### 10.7.1   THE PHYSICAL CHARACTERISTICS OF THE DEVELOPED PROSTHETIC HAND

Selection of sizes and lengths of the phalanx in designing the prosthetic hand was done by referring to the theories introduced by Hamilton and Park in their article (Park et al., 2003). The motion path attained in the experiments was in accordance with the equiangular shape quoted by Park et al. (2003). The design of the prosthetic hand emphasized on mechanism simplicity and robustness. The prototype employed simple pulling tendon mechanism for producing flexion and a resistive torsion spring

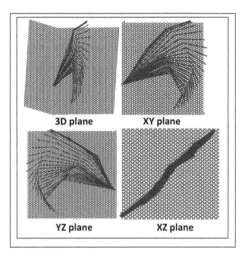

**FIGURE 10.16**    Trajectory of the index finger.

for return mechanism. A similar mechanism was also employed by the famous commercial artificial hand, the Shadow Hand. The structural design of the prosthetic hand developed in this study was made simple for easy assembling and dissembling without destructing any of the components. An added fuzzy controller controlled the supply voltage to the DC motors. Feedback sensors were installed in the prosthetic fingers to improve the quality of the prosthetic hand. The fuzzy controller was able to control the output of the DC motors continuously until the joint reach the input angle, and capable to stop action when the force sensor on the volar surface experienced a force of more than the threshold value.

## 10.7.2 Experimental Results

The main concern was to study the kinematic and kinetic characteristics of the prosthetic hand developed in this study. Experiments were carried out to investigate the performance and robustness of the design. The hand was tested prior to integration with the fuzzy controller. Movement executed by each segment was observed and the motion path was studied. To confirm that the movement angles corresponded to the input value, potentiometers' readings were taken at each joint. The maximum input value of 90° was fixed for this study.

Based on the potentiometer's readings obtained, the angular displacements occurring at each joint were sufficient to support the four essential positions required in this study. The working spaces produced by each joint are between 10–90°. However, in this study, the maximum angle is not crucial since the predetermined essential tasks do not require positioning up to the extreme angle value. Therefore, the prosthetic hand was able to execute the four desired tasks without reaching the maximum angle specified for each joint.

The percentages of error for readings taken at the potentiometers are calculated. The results are shown in Table 10.3. The biggest value of error was recorded at the

**TABLE 10.3**

**Percentages of Error Occurring at Each Joint**

| Fingers | Segments | Percentage of Error (%) |
|---------|----------|-------------------------|
| Index   | DIP      | 8.9                     |
|         | PIP      | 4.4                     |
|         | MCP      | 3.3                     |
| Middle  | DIP      | 22.2                    |
|         | PIP      | 8.9                     |
|         | MCP      | 6.7                     |
| Ring    | DIP      | 10.0                    |
|         | PIP      | 4.4                     |
|         | MCP      | 5.6                     |
| Little  | DIP      | 6.7                     |
|         | PIP      | 6.7                     |
|         | MCP      | 1.1                     |

DIP joint. The possible reason why the DIP joints are more susceptible to error is that distal position required larger torque to achieve the final position and that friction built up from the movement of the proximal segments, thus resulting in premature stopping. Beside the potentiometers, the hand was also equipped with force sensors to measure the contact forces between the object handled and the volar surface of the finger. With the presence of force sensors, the hand can conform to the shape of the objects handled without squashing it.

The hand was further tested with integration of the BCI system. In this mode of the experiment, subjects were not required to key – in the intended flexion angle into the fuzzy controller GUI input box. The input signals were retrieved via electrodes placed on the subject's scalp. The subject was required to perform two or three mental tasks: imaginary right-hand movement, imaginary left-hand movement and imaginary both feet movement. By mean of imaginary movements, the subject was able to control the prosthetic hand according to the objective. The subject, however, must be trained to control the prosthetic hand. The proposed BCI system can control the prosthetic hand during online experiments. Performance of the prosthetic hand in a BCI environment is very much dependent on factors such as the subject's EEG control ability, system performance, feedback delay and 50Hz interference. For the current study, there was no experimental work done to demonstrate the actual performance of the BCI controlling a prosthetic hand. The four tasks focused on the study were basic functional activities commonly performed by the human hand. The selection of the hand tasks can be observed in experiment to investigate the functional strength of the hand (Chao, 1989).

### 10.7.3 Functional Strength

The functional strength on each joint is dependent on the tendon's material. Terelyne string is used as tendon and able to withstand a maximum functional weight of 45kg.

Therefore, the prosthetic hand is capable of handling functional weight more than what is normally required of the human hand. The proof ring method was used to measure the functional weight experienced by each tendon's material. The tension developed within the proof ring resulted in electrical changes in the Wheatstone bridge circuit. The functional strengths experienced by each tendon were measured in term of changes in voltage. The tendon's material at the DIP joint required more torque to flex if compared to the other joints. This is because the distal position received an accumulative torque from the movement of the middle and proximal segments. Hence, the distal joint often faced premature stopping due to the higher torque and friction in the joint.

### 10.7.4  MOTION PATH AND TRAJECTORY

A motion analysis software was used to investigate the working space occupied by each finger. The path of the fingertip trajectory for each prosthetic finger was similar to Guo's trajectory (Guo et al., 1992). Maximum flexion angles for each joint, however, are varied but approached the value suggested by Becker and Thakor (1988). The motions of each finger were captured using three cameras at different angles. Motions from all three cameras were synchronized and digitized at four positions: fingertip, DIP joint, PIP joint and MCP joint. The stability of the movement could be observed from the pattern of the trajectory. In this study, each finger formed an equiangular motion path. The smoothness of the trajectory line demonstrated the stability of each joint during motion, including the absence of wobbling and jerking effects. However, the trajectory of each joint may contain small errors due to errors in digitizing process. To avoid errors, reflective markers should be made prominent compared to the background.

## 10.8  CONCLUSIONS

A prototype prosthetic hand was successfully fabricated and assembled. The hand was designed to perform an equiangular motion path. The prosthetic hand fulfilled the objective of the study and performed within the range of motion predicted by the theory. A total of 16 DOFs obtained is considered sufficient. However, it can be increased for better dexterity. The fingertip and joint trajectories can be improved by increasing the total number of DOFs. This provides a wider range of motion and promotes better manipulative dexterity. However, increased number of DOFs can result in a different and complicated mechanism. Thus, it will add complexity to the design and controller. The major contributors to the total weight of current prosthetic hand are the DC motors and palm. In future, built-in actuators can be considered to replace the DC motors.

The prosthetic hand in this study was developed specifically for BCI application. Application of BCI technology in controlling a prosthetic hand is a promising method to restore communication a "locked-in" patient with the external environment. To date, an experiment to test the ability to control prosthetic hand has been carried out on healthy subjects. The possibility of implementing this technology to a "locked-in" patient is yet to be assessed.

## REFERENCES

Becker, J. C. & Thakor, N. V. (1988). A study of the range of motion of human fingers with application to anthropomorphic designs. *IEEE Transactions on Biomedical Engineering*, 35(2), 110–117.

Birch, G. E. & Mason, S. G. (2000). Brain-computer interface research at the Neil Squire Foundation. *IEEE Transactions on Rehabilitation Engineering*, 8(2), 193–195.

Chao, E. Y. (1989). *Biomechanics of the hand: a basic research study*. World Scientific, Farrer Road, Singapore.

Guo, G., Gruver, W. A., & Qian, X. (1992). A new design for a dexterous robotic hand mechanism. *IEEE Control Systems Magazine*, 12(4), 35–38.

Park, A. E., Fernandez, J. J., Schmedders, K., & Cohen, M. S. (2003). The Fibonacci sequence: relationship to the human hand. *The Journal of Hand Surgery*, 28(1), 157–160.

# 11 The Effect of Prostheses on Standing Stability

## N A Abu Osman and N Arifin
University of Malaya, Kuala Lumpur, Malaysia

## CONTENTS

## 11.1 INTRODUCTION

Lower limb amputations often result from vascular related diseases (such as neuropathy and peripheral vascular disease), trauma, cancer and congenital anomalies (Nielsen, 2007). On the other hand, in some countries with history of recent war, such as Cambodia and Zimbabwe, amputation due to trauma can account for more than 80% (World Health Organisation, 2004). Generally, below-knee (transtibial) and above-knee (transfemoral) amputations are the most common amputation levels, followed by the ankle, hip and knee disarticulations (47%, 31%, 3%, 2%, 1%, respectively) (WHO, 2004). Amputation has been known to not only affect a person physically and psychologically, but also renders a major challenge for the nation (Gitter and Bosker, 2005; Nielsen, 2007). Hence, amputations cause significant implications for increasing the costs of healthcare systems globally, with annual costs of lower extremity amputations in the USA reaching $4.3 billion (Moxey et al., 2011).

Following amputation, one of the rehabilitation goals is to restore the amputee's activities of daily living by reducing the dependency on others and increasing mobility function. One of the essential and basic skills during early rehabilitation training is to control balance during upright standing (Charkhkar et al., 2020). In fact, standing has been reported to be the most frequent indoor activity performed by the unilateral below-knee amputees in comparison to sitting, lying, transitions and other movement-related activities (Bussmann et al., 1998). Maintaining balance, also known as "postural stability", involves the integration of six important components, which are biomechanical constraints, movement strategies (hip and ankle), sensory (visual, somatosensory, vestibular) strategies, orientation in space, control of dynamics and cognitive processing (Horak, 2006). However, this simple task is challenging due to the loss of muscular and skeletal structures, as well as major impairments in both afferent and efferent inputs which are responsible in controlling postural stability (Vanicek et al., 2009).

Often, during upright standing, persons with lower limb amputation are characterised with poor postural stability (Buckley et al., 2002; Vrieling et al., 2008), rely heavily on the intact limb and primarily dependent on visual information (Buckley et al., 2002; Vanicek et al., 2009) during static and dynamic postural stability control. Therefore, amputees exhibit high prevalence for falls and fear of falling compared to age-matched able-bodied individuals (Miller et al., 2003). In addition to the deteriorating postural stability control due to the proprioception loss in individuals with lower limb amputation, several other intrinsic factors were thought to influence the control of stability during upright standing. Findings from previous studies suggested that the reason for amputation (Seth and Lamberg, 2017), length of residual limb (Lenka and Tiberwala, 2007) and level of amputation (Rougier and Bergeau, 2009) are associated with poor stance balance. Although other extrinsic factors such as the type of suspension and socket may alter the control of postural stability, they are yet to be confirmed (Kamali et al., 2013).

Recent advancements in technology have engendered tremendous transformations in the design and materials used to manufacture prosthetic feet. Although the prosthesis allows amputees to perform many activities of daily living, amputation still incites physical and psychological challenges for the amputee. One of the most important elements of a prosthetic device that should be taken into consideration when selecting appropriate ankle-foot prosthesis is the stiffness of the joint. The stiffness of the prosthetic ankle-foot joint is intended to substitute for the loss of muscles and other soft tissues that surround the ankle-foot complex. Interestingly, recent studies suggested that extrinsic factor from the mechanical properties of the prosthetic foot, such as the stiffness, may influence the stability control in anterior-posterior direction among below-knee amputees (Buckley et al., 2002; Nederhand et al., 2012).

Ideally, a prosthetic ankle–foot unit should be prescribed and carefully designed to adjust its mechanical characteristics to the functional needs of the user. This includes the ability of the foot to replicate the biomechanical characteristics of anatomical foot as close as possible (Fergason, 2007; Mason et al., 2011). Nevertheless, to provide general guidelines during the prescription process, prosthetic feet are classified into four primary types according to the motion they permit, mechanical behaviour and physical design (Hafner, 2005; Hofstad et al., 2009). They are: conventional, single-axis, multi-axis and energy storage and return (ESAR) (Figure 11.1). The conventional foot is the basic design, non-articulated solid ankle cushion heel (SACH) which was developed at the University of California in the early 1950s (Michael, 2004). The single-axis (SA) foot permits 15° plantarflexion and 5–7° dorsiflexion with the front bumper substitutes for the gastrocnemius-soleus eccentric contraction and the rear bumper mimics eccentric contraction of the anterior tibialis (Seymour, 2002). The multi-axis foot consists of rubber block which allow dorsiflexion and plantarflexion in the sagittal plane, with additional motions in transverse

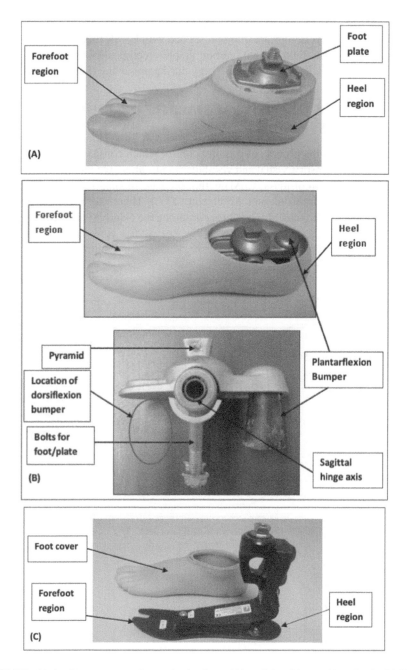

**FIGURE 11.1** Some types of prosthetic feet: (A) solid ankle cushion heel (SACH), (B) single-axis, and (C) energy saving and return (ESAR).

plane such as inversion, eversion and rotation. For amputees leading active lifestyle with high activity levels, they are often prescribed with ESAR foot. The notion of the name is related to its light weight, carbon graphite composites which store forces during loading and release this stored energy during pre-swing (Fergason, 2007).

## 11.2    CONTROL OF POSTURAL STABILITY IN ABLE-BODIED HUMANS

Postural stability is the process of postural control which consist of a complex integration of somatosensory, visual and vestibular inputs along with motor coordination to maintain the centre of mass (CoM) within the base of support (BoS) (Blackburn et al., 2000; Shumway-Cook and Woollacott, 2000). Therefore, it is considered an important aspect in the rehabilitation process among the elderly (Parraca et al., 2011),

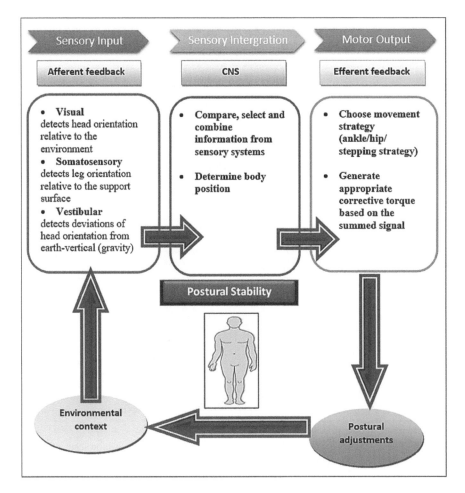

**FIGURE 11.2**    Organization of postural stability control according to systems model approach.

impaired (Salsabili, Bahrpeyma, Forogh and Rajabali, 2011) and amputee (Vrieling et al., 2008; Vanicek et al., 2009) populations. Poor control of postural stability is often associated with the risk of falling which consequently leads to death, injuries and loss of mobility (Winter et al., 1990). The maintenance of stable posture is controlled by the sensory system (vestibular, visual, proprioceptive systems), the central nervous system and musculoskeletal system (Winter et al., 1990). Hence, any deficits of these components will greatly affect the ability to maintain postural stability during standing and walking.

Postural stability control system is viewed based on the systems model which suggest that it is not only dependent on the CNS hierarchical organization from higher and lower level, but also from interrelation between several systems in a concerted manner, as shown in Figure 11.2 (Horak, 2006; Shumway-Cook and Woollacott, 2007). According to the systems model, the maintenance of postural stability involves the integration of six important components which are biomechanical constraints, movement strategies, sensory strategies, orientation in space, control of dynamics and cognitive processing as illustrated in Figure 11.3 (Horak, 2006). Impairments in one or more of the components will lead to postural instability and increase the risk of fall, especially in the elderly and persons with neurological or musculoskeletal disorders.

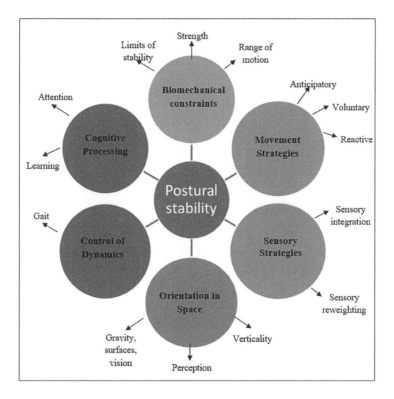

**FIGURE 11.3**   Vital components contributing to postural stability (Horak, 2006).

| Sensory conditions | Descriptions |
|---|---|

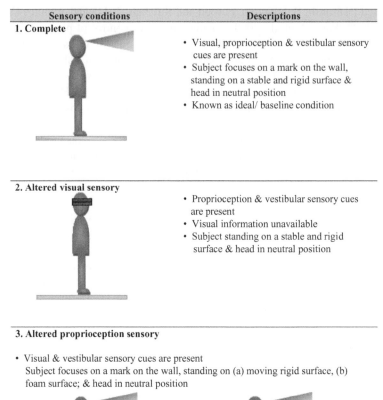

**1. Complete**

- Visual, proprioception & vestibular sensory cues are present
- Subject focuses on a mark on the wall, standing on a stable and rigid surface & head in neutral position
- Known as ideal/ baseline condition

**2. Altered visual sensory**

- Proprioception & vestibular sensory cues are present
- Visual information unavailable
- Subject standing on a stable and rigid surface & head in neutral position

**3. Altered proprioception sensory**

- Visual & vestibular sensory cues are present
  Subject focuses on a mark on the wall, standing on (a) moving rigid surface, (b) foam surface; & head in neutral position

**(a)**              **(b)**

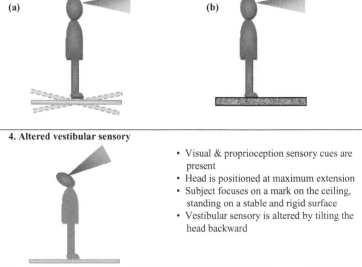

**4. Altered vestibular sensory**

- Visual & proprioception sensory cues are present
- Head is positioned at maximum extension
- Subject focuses on a mark on the ceiling, standing on a stable and rigid surface
- Vestibular sensory is altered by tilting the head backward

**FIGURE 11.4** Different types of sensory alterations conducted in this study for all participants.

Sensory strategies consisting of sensory integration and sensory reweighting are important to maintain stability during altered sensory conditions (for example: eyes closed, visual conflict, compliant surface) as well as during determination of stability limit and vertical position relative to the environment (Mohamadtaghi et al., 2016). Sensory information from the visual, somatosensory and vestibular systems is integrated by the CNS system to appropriately interpret the environment condition (Peterka, 2002; Horak, 2006).

Consequently, these multiple sensory inputs will undergo sensory reweighting process for maintenance of postural stability. For instance, a healthy person walking from a well-lit room to a dark room on a firm surface will rely most on somatosensory system but will change his dependency on visual and vestibular system when standing on an unstable surface. Thus, these unique interactions among the three primary sensory inputs to maintain postural stability has encouraged researchers (Lee et al., 2007; Arifin et al., 2015; Charkhkar et al., 2020) to adopt the sensory organization testing (SOT) for a comprehensive assessment of sensory interactions in balance (Figure 11.4).

The recent version of this test is known as the Clinical Test of Sensory Interaction and Balance (CTSIB), where the subjects are required to stand quietly for 20s under six different conditions which altered the visual and somatosensory information (Lučarević et al., 2020). The theory behind the SOT and CTSIB is that a healthy subject should be able to ignore the inaccurate sensory input and maintain postural stability by utilizing information available from other accurate sensory inputs. Hence, an understanding of the different components and their contribution to the control of postural stability will aid to systematically determine the underlying cause of balance deficit in a particular person (Horak, 2006).

## 11.3 POSTURAL STABILITY CONTROL DURING UPRIGHT STANDING IN LOWER-LIMB AMPUTEES

Individuals with lower limb amputation represent a unique rehabilitation group due to the total loss of neuromuscular and skeletal muscles. Specifically, in a below-knee amputee, the significant role of the ankle joint complex in maintaining balance is jeopardized due to the replacement of the joint complex with prosthetic components with reduced joint mobility and muscle strength (Blackburn et al., 2000; Vanicek et al., 2009). Moreover, it has been shown that an injured ankle joint suffers not only damaged anatomical structures, but also causes deficit to the mechanoreceptors to convey appropriate information to the CNS (Blackburn et al., 2000; Agrawal et al., 2014). As such, it is reasonable to corroborate that a total loss of such joint will give major impact to the ability of controlling postural stability in person with below-knee amputation. In fact, excessive postural instability which may cause falling can occur especially in challenging sensory conditions due to the loss of somatosensory input from the muscles, tendon and skin of the amputated leg (Christie et al., 2020). Hence it is not surprising when community-living lower-limb amputees are reported to have increased risk of falling compared with age-matched, able-bodied people, with 52% of amputees having experienced a fall within a 12-month period (Miller et al., 2001).

Realizing the importance of understanding the reorganization of postural stability in individuals with lower limb amputation, researchers have shown interest in providing evidence-based studies to identify the underlying factors controlling postural stability (Paloski et al., 2006); Previous studies reported that amputees suffer deteriorating balance function due to several influential factors. Although changes in postural control in amputees have been studied, the results have been conflicting. In general, it was accepted that the person with lower-limb amputation exhibited greater body sway than the healthy person (Buckley et al., 2002; Duclos et al., 2009; Jayakaran et al., 2015).

Particularly when compared to a healthy normal group, the below-knee amputee group demonstrated significant increased postural instability in medial-lateral direction during upright standing while looking straight ahead and with eyes closed (Arifin et al., 2015). Similarly, dysvascular elderly amputees with lower scores of circulatory statuses were associated with greater standing instability in medial-lateral direction (Quai, Brauer and Nitz, 2005), whereas Bolger et al. (2014) found conflicting evidence that the CoM displacement and stability margin were similar between individuals with transtibial limb loss and matched controls during randomly applied multi-directional support surface translations. Nevertheless, greater CoP displacement was found in the intact leg during anterior-posterior perturbations, and under the prosthetic leg in medial-lateral perturbations (Bolger et al., 2014).

Cause of amputation is another parameter that can influence standing stability. It has been shown that standing imbalance is more apparent in vascular amputees compared to non-vascular amputees (Molina-Rueda et al., 2017). In contrast, another study showed that postural control in the dysvascular amputation group was not different from the traumatic amputation group in altered sensory conditions (Jayakaran et al., 2015). Additionally, shorter residual limb is associated with poor stance balance due to major loss of proprioceptive sensory at the lower leg (Lenka and Tiberwala, 2010). Regarding the level of amputation, transfemoral amputees exhibit poor balance performance compared to transtibial amputees due to the missing knee joint, combined with greater asymmetry in the loading distribution, and increased deviations of the CoM in the intact leg (Rougier and Bergeau, 2009). Another common consensus from previous studies was that standing load distribution was greater on the intact leg than the amputated leg (Vrieling et al., 2008; Duclos et al., 2009; Hlavackova et al., 2011; Nederhand et al., 2012). Overall, people with lower-limb amputation are faced with poor postural stability control despite the amputation aetiology or level which suggests that the residual limb is not sufficient to completely compensate for the foot as a source for proprioception sensory inputs (Quai et al., 2005).

Although the type of prosthetic knee joint, suspension and socket may influence the standing stability, there is not enough evidence to support such notion (Lenka and Tiberwala, 2010; Kamali et al., 2013). Interestingly, the passive mechanical properties of the prosthetic foot, one of which is the stiffness, have been suggested to influence stability control in anterior–posterior direction among below-knee amputees (Buckley et al., 2002; van der Linde et al., 2004; Curtze et al., 2012; Nederhand et al., 2012; Kamali et al., 2013). In fact, the work of Arifin et al. (2014a, 2014b) showed a reduced body sway in AP and ML direction compared to healthy person which may be due to the stiff ankle of the prosthetic foot. Some researchers suggested the stiffness of prosthetic foot provides an external torque to the knee joint to sustain its stability (Johannesson et al., 2010; Mackenzie et al., 2004).

While these previous studies hypothesized the possible role of prosthetic stiffness in postural stability control, only Nederhand et al. (2012) objectively assessed the relationship between the ankle stiffness of the different prosthetic feet of the amputees and the balance performance during the platform perturbations. From the findings, the prosthetic ankle stiffness was significantly correlated with the dynamic balance control in above- and below-knee amputees. By considering the positive correlation between stiffness and balance control, it may augment the hypothesis that the choice of prosthetic feet with specific properties can influence the postural stability during upright standing. As such, they argued that a stiffer prosthetic foot coupled with balance training could improve the motor skills in utilizing the prosthetic foot as a stabilizing mechanism. However, due to the heterogeneity in terms of amputation level and prosthetic foot types involved in this study, the authors urged for further investigation with repeated testing using different prosthetic feet in one user.

## 11.4 POSTURAL STEADINESS IN BELOW-KNEE AMPUTEES WHEN WEARING DIFFERENT PROSTHETIC FEET DURING VARIOUS SENSORY CONDITIONS

People with lower limb amputations exhibit a higher incidence of falling than the able-bodied, because of the deficits in controlling movements in medial-lateral or anterior-posterior directions (Miller et al., 2001). This is due to the loss of biological ankle joint and a considerable number of muscles in the lower leg, which causes a lack of active ankle torques produced to restore balance in sagittal plane, deficiency in weight-shifting to control balance in frontal plane, and distorted somatosensory input from the amputated side (Charkhkar et al., 2020). Therefore, a prosthetic foot is prescribed to provide passive stability by reducing the amount of body sway regulated at the relatively stiff ankle joint (Buckley et al., 2002). During altered sensory conditions, researchers showed that standing with eyes closed (Barnett et al., 2002; Vanicek et al., 2009) or standing on compliant surface (Kozakova et al., 2009) contributes to the decrease of standing stability in people with lower limb amputation.

Previous studies have assessed the influence of prosthetic foot types in controlling postural stability among transtibial amputees (Arifin et al., 2014a, 2014b, 2015). The researchers have adopted a computed posturography approach to obtain the stability indexes (overall stability index: OSI, anterior/posterior stability index: APSI, and medial/lateral stability index: MLSI), percentage of test time the subject spent in concentric zones and left/right quadrant. In these studies, transtibial amputees were instructed to maintain standing postural stability during various sensory conditions which are firm support surface-eyes opened-head neutral (EO), firm support surface-eyes closed-head neutral (EC), compliant support surface-eyes opened-head neutral (foam) and firm support surface-eyes opened-head extended (HExt).

These studies showed there was a statistically significant difference between sensory conditions in SACH and ESAR foot for all stability indexes (OSI, APSI, and MLSI) (Figure 11.5). Overall, postural instability was the highest during EC

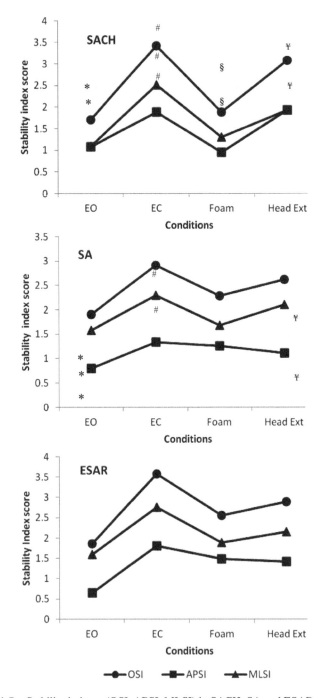

**FIGURE 11.5** Stability indexes (OSI, APSI, MLSI) in SACH, SA and ESAR prosthetic feet during four sensory conditions. Significant differences between two sensory conditions were indicated as *(EO vs EC), ¥(EO vs HExt), #(EC vs foam) and §(foam vs HExt).

compared to other sensory conditions (Buckley et al., 2002; Vanicek et al., 2009, Arifin et al., 2015). In addition, SACH demonstrated significant higher instability for all stability indexes in EC than foam condition as well as in HExt than EO condition. These findings further highlight the importance of visual cues in detecting changes in body orientation with respect to the environment (Shumway-Cook and Woollacott, 2007; Horak, 2006).

Meanwhile, only OSI and APSI in HExt condition were significantly higher than the foam condition. The ESAR foot showed significantly higher OSI and MLSI scores in EC than foam, and significantly higher OSI and APSI scores in HExt condition compared to EO. However, differences in stability indexes were not significant between EO vs. foam and EC vs. HExt conditions for SACH and ESAR feet. As for the SA foot, no significant difference between conditions was observed. Nevertheless, the complete loss of cutaneous, muscle, and joint receptors of the residual limb as well as distorted sensory feedback from the intact limb of the amputees caused inaccurate information, which affected the perceptions and awareness of joint movements and positions (Vanicek et al., 2009; Charkhkar et al., 2020). Therefore, the increase of CoM displacement when amputees were standing on a foam surface was expected. In conditions where the head was extended, the control of postural stability in below-knee amputees was significantly destabilized due to the tilting of otolith organs exceeding its optimal working range (Vuillerme and Rougier, 2005; Lučarević et al., 2020).

The differences in balance indices between sensory conditions were only significant for SACH and ESAR feet, suggesting habitual adaptation to the SA foot for most of the subjects in this study. The insignificant findings for SA foot may also suggest that the subjects utilized the design of single axis mechanical ankle joint that permits movement in the sagittal plane. That is, subjects were able to maintain anterior-posterior and medial-lateral balance by controlling movements at the prosthetic ankle and hip joints. Hence, differences of stability indexes between conditions were not statistically significant for SA foot.

In the context of percentage of time in concentric zones for each sensory condition, amputees demonstrated their ability to sustain CoM excursion within the 0–5° margins, which can be considered as the area of stability. This suggests that constant contractions and relaxations of the muscles in intact and amputated limbs during double-stance quiet standing are well controlled, that the CoM remains close to its zero-centre point in all the tests. In comparison to healthy subjects that spent 85% of time within the 0–5° zone (Arnold and Schmitz, 1998), amputees spent between 82–98% in the same zone. The differences among the four sensory conditions were more significantly apparent in eyes-closed and head extension conditions, suggesting the possible use of these conditions during rehabilitation trainings.

While able-bodied showed 45% and 55% of time spent on right and left quadrant (Arnold and Schmitz, 1998). Transtibial amputees showed 80–94% versus 20–6% loading time on intact and amputated limb, respectively (Arifin et al., 2015). This is consistent with other studies that amputees spent more time and consequently bore more weight on non-affected side than the amputated side (Nadollek et al., 2002; Rusaw, 2019). A possible explanation for this is that the amputee's CoM was located closer to their intact limb than their prosthetic limb during normal standing. Reduced proprioception on the amputated side, due to loss of foot and leg muscles, was also thought to increase the dependency on the intact limb (Christie et al., 2020). Consequently, the asymmetrical loading between intact and amputated leg has been associated with secondary physical conditions such as osteoarthritis on the intact limb, osteoporosis on the amputated limb and back pain (Gailey et al., 2007).

The findings suggest that amputation leads to insufficient control of weight-shifting to maintain an erect posture which caused more instability in medial-lateral direction (Arifin et al., 2020). Other possible explanations include the unwillingness of the amputees to initiate movement at the relatively stiff prosthetic ankle due to lack of confidence, deficit in sensory organization and fear of falling (Miller et al., 2001; Barnett et al., 2012). These results can be used during rehabilitation to determine the direction of the sway to predict fall direction which significantly increases the chances of a hip fracture.

## 11.5  CONCLUSIONS

The maintenance of postural stability involves the integration of sensory systems (visual, proprioceptive and vestibular), central nervous systems and musculoskeletal systems. However, the loss of a biological ankle joint and a considerable number of muscles in the lower leg because of amputation causes distortion in proprioception input which adversely reduces postural stability. Quiet standing with prosthesis is considered a vital skill during the early phase of rehabilitation to achieve independent standing posture before returning to their daily life activities. Previous studies showed that postural instability of amputees commonly occurred during eyes-closed condition, followed by standing with head-extended and standing on foam. Postural stability impairment in below-knee amputees worsens when sensory input was obstructed, despite passive stability provided by the prosthesis. Hence, it is reasonable to suggest that rehabilitation programme should incorporate manipulating of sensory inputs as one of the modalities to improve the ability to maintain postural stability during standing in transtibial amputees.

## REFERENCES

Agrawal, V., Gailey, R.S., Gaunaurd, I.A., O'Toole, C., Finnieston, A., & Tolchin, R. (2014). Comparison of four different categories of prosthetic feet during ramp ambulation in unilateral transtibial amputees. *Prosthetic Orthotic International*. Epub ahead of print.

Arifin, N., Abu Osman, N. A., & Abas, W. A. B. W. (2014). Intrarater test-retest reliability of static and dynamic stability indexes measurement using the Biodex Stability System during unilateral stance. *Journal of Applied Biomechanics*, 30(2), 300–304.

Arifin, N., Abu Osman, N.A., Ali, S., Gholizadeh, H., & Wan Abas, W.A.B. (2014a). Postural stability characteristics of transtibial amputees wearing different prosthetic foot types when standing on various support surfaces. *Scientific World Journal*, 2014, 1–6.

Arifin, N., Abu Osman, N.A., Ali, S., Gholizadeh, H., & Wan Abas, WAB. (2014b). The effects of prosthetic foot type and visual alteration on postural steadiness in below-knee amputees. *Biomedical Engineering OnLine*, 13, 1–10.

Arifin, N., Abu Osman, N.A., Ali, S., Gholizadeh, H., & Wan Abas, W.A.B. (2015). Evaluation of postural steadiness in below-knee amputees when wearing different prosthetic feet during various sensory conditions using the Biodex Stability System (BSS). *Proceedings of the Institution of Mechanical Engineers, Part H: Journal of Engineering in Medicine*, 229(7), 491–498.

Arifin, N., Abu Osman, N.A., & Ali, S. (2020) Stability strategies in transtibial amputees during quiet standing in altered sensory conditions wearing three types of prosthetic feet. *Journal of Mechanics in Medicine and Biology*, 20(2), 1–15.

Arnold, B. & Schmitz, K. (1998). Examination of balance measures produced by the Biodex Stability System. *Journal of Athlete Trainings*, *33*(4), 323–327.

Barnett, C.T., Vanicek, N., & Polma, R.C.J. (2012). Postural responses during volitional and perturbed dynamic balance tasks in new lower limb amputees: A longitudinal study. *Gait & Posture*, *37*(3), 319–325.

Barnett, P. I., McEwen, H. M. J., Auger, D. D., Stone, M. H., Ingham, E., & Fisher, J. (2002). Investigation of wear of knee prostheses in a new displacement/force-controlled simulator. *Proceedings of the Institution of Mechanical Engineers, Part H: Journal of Engineering in Medicine*, 216(1), 51–61.

Blackburn, J.T., Prentice, W.E., Guskiewicz, K.M., & Busby, M.A. (2000). Balance and joint stability: the relative contributions of proprioception and muscular strength. *Journal of Sport Rehabilitation*, 9(4), 315–328.

Bolger, D., Ting, L.H., & Sawers, A. (2014). Individuals with transtibial limb loss use inter-limb force asymmetries to maintain multi-directional reactive balance control. *Clinical Biomechanics*, 29, 1,039–1,047.

Buckley, J.G., O'Driscoll, D., & Bennett, S.J. (2002). Postural sway and active balance performance in highly active lower-limb amputees. *American Journal of Physical Medicine and Rehabilitation*, 81(1), 13–20.

Bussmann, H.B.J., Reuvekamp, P.J., Veltink, P.H., Martens, W.L.J., & Stam, H.J. (1998). Validity and reliability of measurements obtained with an "Activity Monitor" in people with and without a transtibial amputation. *Physical Therapy*, 78(9), 989–998.

Charkhkar, H., Christie, B.P., & Triolo, R.J. (2020). Sensory neuroprosthesis improves postural stability during Sensory Organization Test in lower-limb amputees. *Scientific Reports* 10, 6,984.

Christie, B.P., Charkhkar, H., Shell, C.E. et al. (2020). Ambulatory searching task reveals importance of somatosensation for lower-limb amputees. *Scientific Reports* 10, 10216.

Curtze, C., Hof, A.L., Postema, K., & Otten, B. (2012). The relative contributions of the prosthetic and sound limb to balance control in unilateral transtibial amputees. *Gait & Posture*, 36(2), 276–281.

Duclos, C., Roll, R., Kavounoudias, A., Roll, J.P., & Forget, R. (2009). Vibration-induced post-effects: a means to improve postural asymmetry in lower leg amputees? *Gait & Posture*, 26, 595–602.

Fergason, J. (2007). Prosthetic feet. In M.M. Lusardi & C.C. Nielsen (Eds.), *Orthotics and prosthetics in rehabilitation* (2nd ed.). St. Louis, Missouri: Saunders Elsevier.

Forogh, B., & Rajabali, S. (2011). Dynamic stability training improves standing balance control in neuropathic patients with type 2 diabetes. *Journal of Rehabilitation Research & Development*, 48(7).

Gitter, A. & Bosker, G. (2005). *Upper and lower extremity prosthetics*. (4th ed.). Philadelphia, PA: Lippincott-Raven.

Lučarević, J., Gaunaurd, I., Clemens, S., Belsky, P., Summerton, L., Walkup, M., Wallace, S.P., Yokomizo, L., Pasquina, P., Applegate, E.B., Schubert, M.C., & Gailey, R.S. (2020) The relationship between vestibular sensory integration and prosthetic mobility in community ambulators with unilateral lower limb amputation. *Physical Therapy* 100(8):1,333–1,342.

Hafner, B. J. (2005). Clinical prescription and use of prosthetic foot and ankle mechanisms: a review of the literature. *JPO: Journal of Prosthetics and Orthotics*, 17(4), S5–S11.

Hlavackova, P., Franco, C., Diot, B., & Vuillerme, N. (2011). Contribution of each leg to the control of unperturbed bipedal stance in lower limb amputees: new insights using entropy. *PLoS One*, 6(5), 1–4.

Hofstad, C.J., van der Linde, H., van Limbeek, J., & Postema, K. (2009). Prescription of prosthetic ankle-foot mechanisms after lower limb amputation. *Cochrane Database Systematic Review*, 1, CD003978.

Horak, F.B. (2006). Postural orientation and equilibrium: what do we need to know about neural control of balance to prevent falls? *Age & Ageing*, 35, ii7–11.

Jayakaran, P., Johnson, G.M., & Sullivan, S.J. (2015). Postural control in response to altered sensory conditions in persons with dysvascular and traumatic transtibial amputation. *Archives of Physical Medicine and Rehabilitation*, 96, 331–339.

Johannesson, A., Larsson, G.U., Ramstrand, N., Lauge-Pedersen, H., Wagner, P., & Atroshi, I. (2010). Outcomes of a standardized surgical and rehabilitation program in transtibial amputation for peripheral vascular disease: a prospective cohort study. *American Journal of Physical Medicine and Rehabilitation*, 89(4), 293–303.

Kamali, M., Karimi, M.T., Eshraghi, A., & Omar, H. (2013). Influential factors in stability of lower-limb amputees. *American Journal of Physical Medicine and Rehabilitation*, 92, 1–9.

Knobloch, L. A., Gailey, D., Azer, S., Johnston, W. M., Clelland, N., & Kerby, R. E. (2007). Bond strengths of one-and two-step self-etch adhesive systems. *The Journal of Prosthetic Dentistry*, 97(4), 216–222.

Lee, M.Y., Lin, C.F., & Soon, K.S. (2007). Balance control enhancement using sub-sensory stimulation and visual-auditory biofeedback strategies for amputee subjects. *Prosthetics and Orthotics International*, 31, 342–352.

Lenka, P., & Tiberwala, D. N. (2010). Effect of stump length on postural steadiness during quiet stance in unilateral trans-tibial amputee. *Age (Years)*, 3(1), 50–57.

Lenka, P. K., & Tiberwala, D. N. (2007). Effect of stump length on postural steadiness during quiet stance in unilateral trans-tibial amputee. *Al Ameen Journal of Medical Sciences, 3*, 50–57.

Mackenzie, E.J., Bosse, M.J., Casttillo, R.C., Smith, D.G., Webb, L.X. Kellam, J.F., …, McCarthy, M.L. (2004). Functional outcomes following trauma-related lower-extremity amputation. *Journal of Bone & Joint Surgery*, 86(8), 1,636–1,645.

Mason, Z.D., Pearlman, J., Cooper, R.A., & Laferrier, J.Z. (2011). Comparison of prosthetic feet prescribed to active individuals using ISO standards. *Prosthetics and Orthotics International*, 35(4), 418–424.

Michael, J.W. (2004). Prosthetic suspensions and components. In D.G. Smith, J.W. Michael & J.H. Bowker (Eds.), *Atlas of amputation and limb deficiency: surgical, prosthetic and rehabilitation principles* (3rd ed.). Rosemont, IL: American Academy of Orthopaedic Surgeons.

Miller, W.C., Deathe, A.B., & Speechley, M. (2003). Psychometric properties of the Activities Specific Balance Confidence Scale among individuals with a lower-limb amputation. *Archives of Physical Medicine and Rehabilitation*, 84, 656–661.

Miller, W. C., Speechley, M., & Deathe, B. (2001). The prevalence and risk factors of falling and fear of falling among lower extremity amputees. *Archives of Physical Medicine and Rehabilitation*, 82(8), 1,031–1,037.

Mohamadtaghi, B.B., Hejazi Dinan, P., & Shamsipour Dehkordi, P. (2016). Effect of the selected balance program on postural control of amputees under manipulation of visual, vestibular and proprioceptive systems. *Iranian Journal of War and Public Health*. 8(1), 1–8.

Molina-Rueda, F., Molero-Sánchez, A., Carratalá-Tejada, M., Cuesta-Gómez, A., Miangolarra-Page, J.C., & Alguacil-Diego, I.M. (2017). Limits of stability in patients with vascular (due to diabetes) and nonvascular unilateral transtibial amputation: a cross-sectional study. *International Journal of Rehabilitation Research*, 40(3), 227–231.

Moxey, P.W., Gogalniceanu, P., Hinchliffe, R.J., Loftus, I.M., Jones, K.J., Thompson, M.M., & Holt, P.J. (2011). Lower extremity amputations – a review of global variability in incidence. *Diabetes Care*, 28, 1,144–1,153.

Nadollek, H., Brauer, S., & Isles R. (2002). Outcomes after transtibial amputation: the relationship between quiet stance ability, strength of hip abductor muscles and gait. *Physiotherapy Research International*, 7, 203–214.

Nederhand, M.J., Van Asseldonk, E.H.F., Der Kooij, H.V., & Rietman, H.S. (2012). Dynamic Balance Control (DBC) in lower leg amputee subjects: contribution of the regulatory activity of the prosthesis side. *Clinical Biomechanics*, 27(1), 40–45.

Nielsen, C.C. (2007). Etiology of amputation. In M.M. Lusardi & C.C. Nielsen (Eds.), *Orthotics and prosthetics in rehabilitation*(2nd ed.). St. Louis, MO: Saunders Elsevier.

Paloski, W.H., Wood, S.J., Feiveson, A.H., Black, F.O., Hwang, E.Y., & Reschke, M.F. (2006). Destabilization of human balance control by static and dynamic head tilts. *Gait & Posture*, 23(3), 315–323.

Parraca, J. A., Olivares Sánchez-Toledo, P. R., Carbonell Baeza, A., Aparicio García-Molina, V. A., Adsuar Sala, J. C., & Gusi Fuertes, N. (2011). Test-Retest reliability of Biodex Balance SD on physically active old people.

Peterka, R.J. (2002). sensorimotor integration in human postural control. *Journal of Neurophysiology*, 88, 1,097–1,118.

Quai, T. M., Brauer, S. G., & Nitz, J. C. (2005). Somatosensation, circulation and stance balance in elderly dysvascular transtibial amputees. *Clinical Rehabilitation*, 19(6), 668–676.

Rougier, P.R. & Bergeau, J. (2009). Biomechanical analysis of postural control of persons with transtibial or transfemoral amputation. *American Journal of Physical Medicine and Rehabilitation*, 88, 896–903.

Rusaw, D.F. (2019). Adaptations from the prosthetic and intact limb during standing on a sway-referenced support surface for transtibial prosthesis users. *Disability and Rehabilitation: Assistive Technology*, 14(7), 682–691.

Seth, M. & Lamberg, E. 2017. Standing balance in people with transtibial amputation due to vascular causes: a literature review. *Prosthetics and Orthotics International*, 41(4), 345–355.

Seymour, R. (2002). *Prosthetics and orthotics: lower limb and spinal*. Philadelphia, PA: Lippincott Williams & Wilkins.

Shumway-Cook, A., & Woollacott, M. H. (2007). *Motor control: translating research into clinical practice*. Philadelphia, United States: Lippincott Williams & Wilkins.

Shumway-Cook, A., & Woollacott, M. H. (2000). Attentional demands and postural control: The effect of sensory context. *Journal of Gerontology*, 55A(1), M10–M16.

van der Linde, H., Hofstad, C.J., Geurts, A.C., Postema, K., Geertzen, J.H. & van Limbeek, J. (2004). A systematic literature review of the effect of different prosthetic components on human functioning with a lower-limb prosthesis. *Journal of Rehabilitation Research and Development*, 41(4), 555–570.

Vanicek, N., Strike, S., McNaughton, L., & Polman, R. (2009). Postural responses to dynamic perturbations in amputee fallers versus nonfallers: a comparative study with able-bodied subjects. *Archive of Physical Medicine and Rehabilitation*, 90(6), 1,018–1,025.

Vrieling, A.H., van Keeken, H.G., Schoppen, T., Otten, E., Hof, A.L., Halbertsma, J.P., & Postema, K. (2008). Balance control on a moving platform in unilateral lower limb amputees. *Gait & Posture*, 28(2), 222–228.

Vuillerme, N. & Rougier, P. (2005). Effects of head extension on undisturbed upright stance control in humans. *Gait & Posture*, 21, 318–325.

Winter, D.A., Patla, A.E., & Frank, J.S. (1990). Assessment of balance control in humans. *Medical Progress Through Technology*, 16, 31–51.

World Health Organisation Staff, & World Health Organization. (2004). *Laboratory Biosafety Manual*. World Health Organization.

# 12 Bamboo as a Pylon Material

## N A Abu Osman and H N Shasmin

University of Malaya, Kuala Lumpur, Malaysia

## CONTENTS

## 12.1 INTRODUCTION

A transtibial amputation is an amputation that takes place between the ankle and knee joints. The transtibial prosthesis is made up of three main components: a socket, a pylon and a foot–ankle assembly. The anatomical shank is represented by the pylon part, which is used to attach the socket to the foot. Transtibial and transfemoral amputees also have a range of completely functional prosthetic options. Amputated athletes, for example, have an equal number of choices in the market to suit sprinting, skiing, golf, swimming and other sports. With high-tech prostheses, it can cost several thousand dollars in Western countries and support only a small number of patients in Malaysia. Since most occupations in the country require some form of physical labour,

DOI: 10.1201/9781003196730-12

having accessible prosthetic legs is critical for any amputee. The research takes a novel approach to lowering transtibial prosthetic costs by developing an affordable pylon from a natural source: bamboo, a fibre-reinforced composite material with favourable mechanical properties that warrants its use as a structural material.

## 12.2   BAMBOO

Bamboo is a one-of-a-kind plant that is known as "the poor's wood" in India, "the people's mate" in China, and "the brother" in Vietnam. Bamboo is a composite material composed of fibres and cellulose. Bamboo plants have been used for a variety of purposes for thousands of years. Bamboo has been used in a wide range of goods, from household items to industrial applications. Toys, food containers, furniture, flooring material, handicrafts, pulp and vessels, charcoal, paper, firearms, and musical instruments, are some examples of bamboo items. The outstanding mechanical behaviour of bamboo in countless everyday applications has shown its great potential for use as a supporting material for amputees. Certain bamboos provide the only suitable material that is sufficiently cheap and plentiful to satisfy the need for eco-material pylon in many overcrowded tropical regions. According to research, the physical and mechanical properties of bamboo are primarily determined by the species, soil, silvicultural treatment, harvesting season, felling age, and moisture content (Jain et al., 1992; Lee et al., 1994; Sulaiman et al., 2006). Shin et al. (1989) studied the flexural, compressive, inter-laminar and tensile, shear properties of bamboo epoxy composites with different lamina numbers. The bamboo evaluated in this study belonged to the genus *bambusa paravariabilis*, which is laminated into three, five and seven layers. These samples were then put through their paces on a Universal Testing Machine (UTM) with a crosshead speed of 1mm/min.

## 12.3   METHODOLOGY

### 12.3.1   DEVELOPMENT OF BAMBOO PYLON PROSTHESIS

Harvesting, culm drying, bamboo treatment and bamboo lamination are the four stages in the bamboo pylon process. This study selected sympodial rhizomes bamboo, *bambusa heterostachya*, to produce a bamboo pylon because this species is widely and abundantly found in Malaysia. Furthermore, the species' physical characteristics make it an outstanding option for sensible work. Furthermore, the fully grown *bambusa heterostachya* was well suited for pylon development.

### 12.3.2   HARVESTING LOCAL BAMBOO

Between August 20th and 30th, 2006, sound internodes of matured bamboo were harvested from bambusetum at the Forest Research Institute Malaysia (FRIM). Maturity of the culm was needed for the best use of bamboo when harvesting. Bamboo, unlike wood, lacks the vascular cambium layer (Nghia, 2006; Xiaobing, 2007), so there is no marking of growth diameter and, as a result, no year rings.

Thus, to determine the maturity of the culm (3–4 years), this study used Fu's (2001) methods of judging the colour of the culm and calculating the volume of the bamboo clump from a diameter of 5–3 metres. After selecting the perfect bamboo, it was removed from a clump by cutting it just above a node about 20cm above the ground. The bottom portion of three-year-old bamboo was used to prepare the samples. The length and diameter of each chosen culm were the same as the traditional pylon dimensions, which were approximately 30mm and 200mm, respectively (outer diameter). The appropriate culm in a bamboo pole was selected when the inner diameter was half that of the outer diameter, which was normally at the fourth and fifth internodes or 2/3 of the bamboo length from the ground level of *bambusa heterostachya*.

### 12.3.3 Culm Drying

The culms were handled as soon as they were removed. The adequate preservation of the culm was a critical factor. The well-shaped samples were oven-dried to extract the water material, preventing fungus growth and culm decomposition. Because of the high concentration of starch, bamboo was extremely vulnerable to attack by staining fungi and powder-post beetles (Mathew and Nair, 1990). The drying process was carried out for 72 hours in a 200 °C oven.

### 12.3.4 Oil Treatment

After that, the bamboo culms were cut into specific sizes and pre-treated with hot oil soaking. As recommended by FRIM, palm oil (V-sawit oil) was used in this analysis. The bamboo was treated with oil using an electrical oil-curing machine. Since it was organic and had a high boiling point, palm oil was used as the heating medium. First heated the palm oil at temperature of 60°C. The bamboos were then immersed in the heated oil by enclosing them in a metallic encasement. After 30–90 minutes, the bamboo was removed at 120–180°C.

### 12.3.5 Lamination

It is important to understand the culm's technical properties as well as its low resistance to pathogens. At least three layers of coating were added to the culms' exterior surface. These external coatings will provide an aesthetic effect to the culms while also acting as a defensive layer against fungus or mite infestation, which could later weaken the mechanical integrity of the pylon. As a result, the durability of the bamboo pylon produced by the process was significantly enhanced. One layer of vinyl urethane adhesive with polyvinyl acetate as hardener was added tangentially to the outer and inner surfaces of the culm for lamination. The culms were dried for 24 hours after the coating layer was applied. This phase would harden the coating materials before adding another layer until there were three laminated layers, each about 1mm thick (Figure 12.1). The culms were ready for mechanical testing after the lamination layers dried.

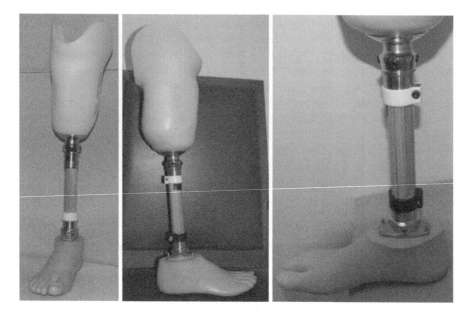

**FIGURE 12.1**   The final product of bamboo pylon.

## 12.4   MECHANICAL PROPERTY TESTS

At the age of 3–4 years, 50 bamboo culms of *bambusa heterostachya* were collected for mechanical property analysis. This bamboo had an average culm length of 5.49m, a culm diameter of about 3.5cm at the bottom, and a diameter of about 2.3cm at the top. The culm wall's average thickness was 0.97cm. Ten samples from each culm 1 to culm 5 were chosen for the experiments. The research included three mechanical tests: flexural tensile, bending tensile, and compression. Based on the American Society for Testing and Materials (ASTM) Standards, all the samples' dimensions and shapes were cut based on the assigned mechanical test.

## 12.5   CLINICAL TESTS

Subjects for the study were selected from those who wore a similar part of a transtibial prosthesis, such as a PTB socket with liner, a stainless-steel pylon and a SACH foot. Four weeks before the clinical assessments, each subject received a new socket and a SACH foot to fit the bamboo pylon. The amputee subjects were fitted with the new transtibial prosthetic leg at the University Malaya Medical Centre's Prosthetics and Orthotics Workshop (UMMC). A new range of prostheses was manufactured and fitted by the same Category 2 prosthetist using normal procedures. Both clinical evaluations were carried out at the University of Malaya's Motion Analysis Laboratory, Department of Biomedical Engineering. The gait analysis method and socket/stump interface pressure analysis were both used. The 3D biomechanics of a participant's lower limb were captured for the gait analysis test using a commercial motion capture coordination known as the VICON motion capture system® combined with

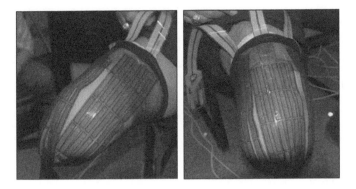

**FIGURE 12.2** Sensor installation in an F-socket. It is secured by a skin stocking which protects the anterior, dorsal, vertical and lateral sections of the subject's residual limb.

Nexus 1.3 software and two KISTLER® force plates. This study's motion capture system (Model FX-20) consisted high-speed video cameras and of six advanced infrareds. After the subject had put on the prosthetic leg, sixteen 14mm motion reflective markers were mounted on the subject's lower limb body to mark bone segments. The pressure distribution and forces acting on the interface between the residual limb and the prosthetic socket were provided by Tekscan Inc.'s F-socket system®. The calibrated F-socket sensors were positioned on the residual limb to measure pressure from all edges: medial, posterior, anterior, and lateral (Figure 12.2). The benefit of placing the sensors on the subject's stump rather than the inner of the hard socket was that they did not need to be taken off and could be used again when the subject switched to the other pair of prosthetic legs. The Medical Ethics Committee at UMMC accepted the bamboo pylon clinical review. Before taking part in the study, all participants signed written consent forms after being granted a verbal and written description of the study's procedure.

## 12.6 RESULTS AND DISCUSSION: MECHANICAL STRENGTH TEST

Fifty bamboo pylon samples were tested in three-point bending tests, tensile and compression. The nodes were removed from all the specimens included in the analysis. Bamboo pylon specimens had a total specific gravity of 0.52 and a moisture content of 12%. In the lateral, tangential and longitudinal directions, the average percent shrinkage of the specimens from green to bamboo pylon was 18.21%, 9.25%, and 0.02%, respectively. Table 12.1 displays the outcomes of mechanical experiments.

The bamboo pylon's mean strength was 230.3MPa (SD ± 6.2MPa). The bamboo pylon's modulus tensile, $E_t$, was 49.5GPa (SD 6.9 GPa). Figure 12.3 depicts tensile specimens tested in accordance with ASTM–D 143–94. Tensile loads rose linearly with increasing strain until the point of ultimate load, when bamboo fibres broke and shown brittle fracture. Beyond the rupture points, the stress-strain curves showed sharp, phased decreases. Tensile fracturing of bamboo was mostly caused by longitudinal cracking in the same direction as the fibres. According to the tensile curves, bamboo was a fragile fibre with no strain-hardening property. According to Joachim et al. (2007), the

## TABLE 12.1
## Bamboo Pylon Mechanical Properties

| Mechanical Properties | Units | Bamboo Pylon |
|---|---|---|
| Tensile Yield | $\sigma_t$, MPa | $230.3 \pm 6.2$ |
| Tensile Modulus | $E_t$, GPa | $49.5 \pm 6.9$ |
| Compression Yield | $\sigma_c$, MPa | $132.6 \pm 3.3$ |
| Compressive Modulus | $E_c$, GPa | $30.7 \pm 4.7$ |
| Flexural Yield | $\sigma_b$, MPa | $220.6 \pm 3.5$ |
| Flexural Modulus | $E_b$, GPa | $27.2 \pm 4.5$ |

No. samples, N = 50.
Values are mean ± standard deviation.
No significant difference for both groups (p >0.05).

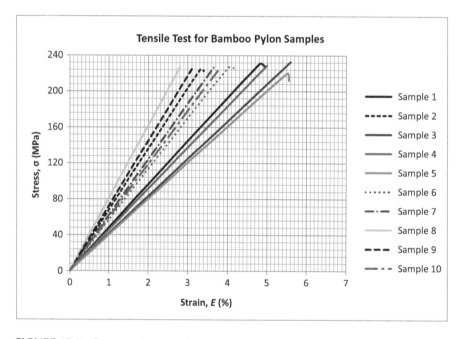

**FIGURE 12.3**   Stress-strain curves for ten bamboo pylon samples under tensile load.

breaking strength and ultimate strength of one brittle substance is comparable. Tensile strength was usually negligible in this kind of material, and the material was mainly used for other types of mechanical strength.

ASTM–D 3410 norms were used to investigate the compressive properties. To ensure that cracks occurred in the centre of the test laminate, the two ends of a sample were clamped with 30mm (SD ± 4.5mm) pylon adapters. The sample arranged to the Universal Testing Machine using a pair of longitudinal strain gauges mounted to the ends, and the angle was changed so the gauges displayed the same strain to ensure the absence of bending moment. For modulus and compressive strength, $\sigma_c$ in bamboo pylon, $E_c$ were 30.7GPa (SD ± 4.7GPa) and 132.6MPa (SD ± 3.3MPa), respectively. About the fact that the compressive intensity of the bamboo pylon was lower than the tensile strength, the stress-strain curves revealed that it had better physical properties under compression loads, with the appearance of a strain-hardening field (Figure 12.4).

Because of its plastic-deformity property, a substance with a strain-hardening area would not shatter at its ultimate power (Verterra, 2007). This demonstrates that when applied under compression load, bamboo behaves similarly to steel and other ductile materials, but behaves differently when applied under tensile load. The distinction was potentially that bamboo, unlike synthetic materials such as titanium and stainless steel, is not a homogeneous substance. Because of their homogeneous nature, these industrial materials can have similar strength under compressive and tensile stress.

The 3-point bending test was carried out in accordance with ASTM–D 3043–95 requirements. For bamboo pylon, the flexural modulus, Eb, was 27.2GPa (SD 4.5GPa), and the mean flexural pressure, b, was 220.6MPa (SD 3.5MPa). Figure 12.5 depicts the stress-strain curves for all types of 3-point bending tests. The centre of the sample was localized bending fractures where the load was applied. In this analysis, the stress-strain curve revealed an early linear section, followed by a non-linear segment above 50% of fracture load before ultimate strength was applied. Subsequently there was an unusual phased decrease in load.

Hot oil was used to treat the bamboo pylon. The oil treatment is applied to different types of wood to increase their dimensional stability and resistance to

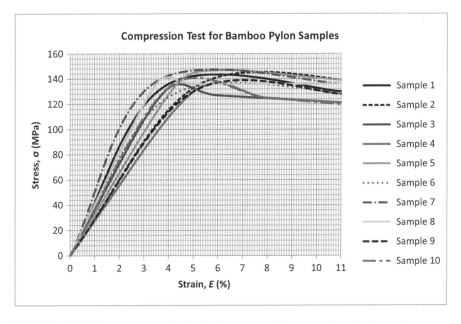

**FIGURE 12.4**  Under compression load, the stress-strain curves of ten bamboo pylon samples were plotted.

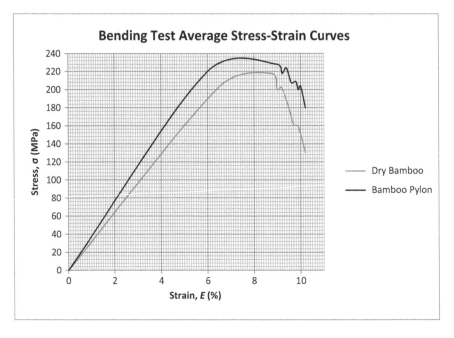

**FIGURE 12.5** Stress-strain curves for dried bamboo and bamboo pylon 3-point bending tests. The graphs reflect the average of 50 samples (N = 50).

bio-deterioration. Various forms of oil treatment processes were studied, and it was discovered that process factors such as wet or dry operation, types of oils and temperatures were major parameters determining the total properties of the treated materials. The treated oil entered both the cell wall and the cell lumen. Oil entry into the cell wall and lumen interacted with the interfacial surface of bamboo, reducing adhesion. To stop this situation, the bamboo pylon for the transtibial prosthesis was handled at 120–180°C for less than 90 minutes. In traditional transtibial pylons, stainless steel and aluminium are often used. Aluminium has a mechanical power of 48MPa in its annealed form (Shasmin et al., 2008). Another new composite used in prosthetic pylons is fibre-reinforced plastic, which has a Young's modulus of 12.23GPa (Hahl et al., 2000). As compared to traditional materials, bamboo is expected to be a good substitute for the current pylon material in transtibial prosthesis. Bamboo pylon is three times stronger than fibre-reinforced plastic and two times stronger than aluminium, with yield compressive stress and Young's modulus of 132.6MPa and 30.7GPa, respectively. Figure 12.6 compares the Young's modulus of different materials to that of bamboo pylon.

## 12.7 CLINICAL TEST

### 12.7.1 Subjects

Five amputees from the Department of Rehabilitation Medicine at UMMC were recruited for the proven trials. The enrolled amputees had to be stable transtibial

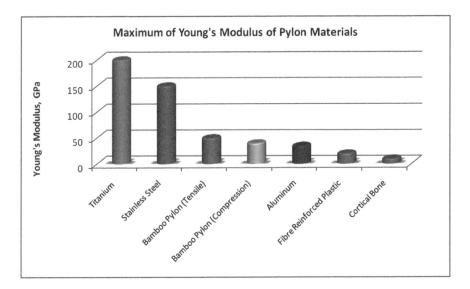

**FIGURE 12.6** Young's modulus with traditional pylon fabrics, bamboo pylons and bone. Young's Modulus of bamboo pylon is greater than that of aluminium and fibre-reinforced plastic. Reproduced from Rho (1993).

amputees who could walk individually for at least 15 metres without problems. Amputees with neurological or orthopaedic impairments, as well as those walking with the assistance of a gait assistance, were disqualified from the trial. Table 12.2 summarizes the demographic information for the topics.

### 12.7.2 Spatio–temporal Parameters

Spatio–temporal parameter analyses were divided into four components: prosthetic and sound side from both prosthetic legs. Each subject's data was summed over at least three gait periods. The statistical test was run using one-way ANOVA, with the significance degree set at 0.05.

---

**TABLE 12.2**

**Parameter's Amputees**

| Subject | Age (years) | Body Mass (kg) | Height (m) | Years since Amputation | Amputated Side | Reason for Amputation | Prosthetic Shank Length (cm) |
|---|---|---|---|---|---|---|---|
| S1 | 35 | 108 | 1.76 | 1 | Right | Diabetes | 25.4 |
| S2 | 62 | 55 | 1.49 | 4 | Right | Diabetes | 19.3 |
| S3 | 57 | 86 | 1.72 | 5 | Right | Diabetes | 23.7 |
| S4 | 45 | 79 | 1.57 | 4 | Right | Diabetes | 22.1 |
| S5 | 64 | 100 | 1.63 | 7 | Right | Diabetes | 24.5 |

## TABLE 12.3
### Temporospatial Data for Subjects

|  |  | BPL | | SPL | |
| --- | --- | --- | --- | --- | --- |
|  |  | Prosthetic | | Prosthetic | |
| Variables | Units | Side | Sound Side | Side | Sound Side |
| Cadence | steps/min | 73.59 ± 4.55 | 76.79 ± 4.33 | 76.80 ± 6.91 | 81.52 ± 9.05 |
| Step Length | m | 0.38 ± 0.08 | 0.37 ± 0.16 | 0.40 ± 0.14 | 0.47 ± 0.17 |
| Step Width | m | 0.25 ± 0.04 | 0.24 ± 0.04 | 0.33 ± 0.10* | 0.33 ± 0.12* |
| Walking Speed | m/s | 0.45 ± 0.13 | 0.48 ± 0.13 | 0.57 ± 0.20 | 0.59 ± 0.18 |

Values are mean ± standard deviation from (N = 6).
* Significant difference (p <0.05) between category (BPL and SPL).
No significant difference was recorded between legs within same category (p >0.05).

There was no substantial difference (p >0.05) observed between bamboo prosthetic leg (BPL) and stainless-steel prosthetic leg for the spatio–temporal parameters tabulated in Table 12.3. (SPL). Values on all sound sides were marginally higher than those for the prosthetic side, except for step length and step width for BPL, which were significantly higher than those for the sound side. The stage width differed significantly between the BPL and SPL categories (p = 0.01). BPL had equal cadence, step length, and walking speed to SPL (p >0.05). All values obtained while walking with SPL were greater than those obtained while walking with BPL. The box plots for cadence are seen in Figure 12.7. Cadence for the sound side with SPL had the highest value, while cadence for the prosthetic side with BPL had the lowest. The cadence frequency of the sound side of BPL and the prosthetic side of SPL is almost identical.

It has not been shown that the use of lightweight prostheses benefits amputees. None of the previous studies (Godfrey et al., 1977; Skinner and Mote, 1989; Hale, 1990; Czerniecki et al., 1994) a study of the impact of prosthetic weight on gait in amputees found that a lighter prosthesis improved subject preference, walking pace and energy expenditure. According to Meikle et al. (2003), short-term operation with improved prosthetic mass has no substantial adverse impact on gait rhythm. As a result, there was no difference in spatio–temporal parameters found when subjects wore BPL instead of SPL during the gait study.

### 12.7.3 GROUND REACTION FORCES

Even though GRF were obtained in all three directions, only the vertical and A/P directions were studied. Figure 12.8 contrasts the prosthetic and sound side mean vertical and A/P GRF curves for one subject (S3) using both the BPL and SPL, standardized by body weight. As S3 walked with SPL, the vertical GRF of the sound side developed a classic curve (Figure 12.8), similar to when a regular subject walked barefoot (Whittle, 1996), and the prosthetic side produced an asymmetrical curve from the time of initial touch to the loading process (Figure 12.9). Furthermore, when S3 walked with BPL, the vertical GRF on the sound leg was asymmetrical,

## TABLE 12.4
## The BPL and SPL Used for Mean ± Standard Deviation of GRFs for Six Subjects During Walking

| | | BPL Prosthetic Side | BPL Sound Side | SPL Prosthetic Side | SPL Sound Side |
|---|---|---|---|---|---|
| **GRF (N/kg)** | | | | | |
| Vertical | $Fz_1$ | 1.05 ± 0.45 | 1.13 ± 0.34 | 1.17 ± 0.31 | 1.02 ± 0.33 |
| | $Fz_2$ | 1.01 ± 0.67 | 0.97 ± 0.52 | 0.89 ± 0.50 | 0.87 ± 0.45 |
| | $Fz_3$ | 0.95 ± 0.34 | 1.06 ± 0.23 | 1.11 ± 0.32 | 1.01 ± 0.30 |
| A/P | $Fx_1$ | −0.20 ± 0.03 | −0.23 ± 0.05 | −0.43 ± 0.01* | −0.22 ± 0.01 |
| | $Fx_2$ | 0.43 ± 0.07 | 0.31 ± 0.04 | 0.47 ± 0.03 | 0.26 ± 0.02 |
| **Range of Motion (°)** | | | | | |
| Hip Angle | Maximum Extension | 5.25 ± 0.26 | 0.02 ± 0.00* | 4.79 ± 0.24 | 14.17 ± 0.71* |
| | Maximum Flexion | 37.97 ± 1.90 | 32.12 ± 1.65 | 29.27 ± 1.45 | 28.75 ± 1.45 |
| Knee Angle | Maximum Extension | 10.00 ± 0.50 | 7.38 ± 0.21 | 8.92 ± 0.45 | 10.15 ± 0.51 |
| | Maximum Flexion | 60.26 ± 3.05 | 58.41 ± 2.92 | 45.68 ± 2.28 | 39.19 ± 1.96 |
| Ankle Angle | Maximum Plantar Flexion | 17.93 ± 0.89 | 19.80 ± 0.99 | 28.44 ± 1.42* | 19.80 ± 0.99 |
| | Maximum Dorsiflexion | 4.60 ± 0.23 | 3.92 ± 0.09 | 4.84 ± 0.24 | 4.02 ± 0.20 |

Values are mean ± standard deviation from (N = 6).
* Significant difference (p <0.05).

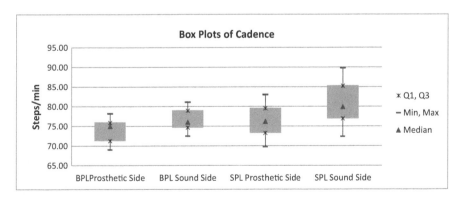

**FIGURE 12.7**   Box plots of subjects' cadence.

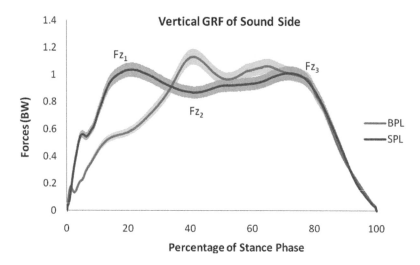

**FIGURE 12.8** Vertical GRF of sound side for one subject walking at a self-selected tempo using the BPL and SPL. Body weight was used to normalize GRFs (BW). The shaded field denotes the standard deviation.

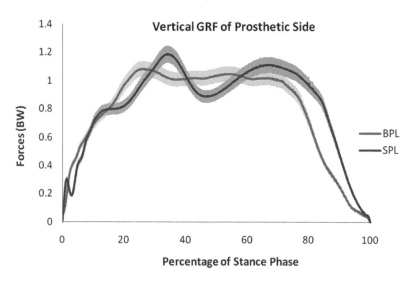

**FIGURE 12.9** Vertical GRF of prosthetic side for one subject walking at self-selected pace using the BPL and SPL. Body weight was used to normalize GRFs (BW). The shaded field denotes the standard deviation.

while the prosthetic leg produced a flat curve from loading to terminal stance step. When walking with BPL and SPL, the A/P GRF for S3 showed almost equal curves on the sound line (Figure 12.10). When walking with BPL, though, breaking force was weaker on the prosthetic side than when walking with SPL. Both prosthetic legs had similar propulsion powers on the prosthetic limb (Figure 12.11). These discoveries were made by both subjects in the same way (Table 12.5).

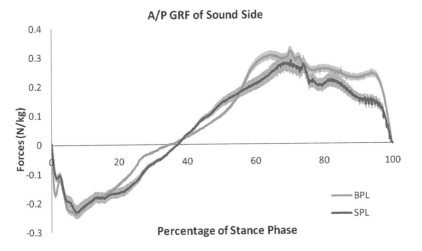

**FIGURE 12.10** A/P GRF of the sound side for one subject walking at a self-selected tempo using the BPL and SPL. Body weight was used to normalize GRFs (BW). The shaded field denotes the standard deviation.

The maximum force amplitude was described as the highest peak on the force curve.

Three maximal forces reflect vertical GRF (Figure 12.8). During the loading process, Fz1 was the maximum vertical power. Fz2 represented vertical force in midstance, while Fz3 represented full vertical force in late terminal stance. For research purposes, two forces were defined in A/P GRF: Fx1 as the maximum breaking force and Fx2 as the maximum propulsion force (Figure 12.11). The time is estimated in seconds and represents the time from the beginning of the stride to the moment the force exceeds the maximum amplitude shown by the peak. Both force measurements were normalized to body mass. The amplitude of the first peak of the vertical GRF usually shows the locomotor system's ability to withstand shock at gait. Previous research on people who had unilateral transtibial amputations found that the first peak of the vertical GRF on the prosthetic limb was 5–15% greater than the sound

**TABLE 12.5**

**Mean ± Standard Deviation of Residual Limb Interface Pressure Was Calculated for Six Participants when Walking with BPL and SPL**

|  | BPL | SPL |
|---|---|---|
| **Maximum pressure (kPa)** | | |
| Anterior | 61.34 ± 32.23 | 62.31 ± 44.26 |
| Posterior | 31.14 ± 17.81 | 60.64 ± 14.53* |
| Medial | 55.16 ± 26.55 | 63.75 ± 31.89 |
| Lateral | 51.96 ± 15.32 | 59.06 ± 15.43 |

Values are mean ± standard deviation.

* Significant difference (p <0.05).

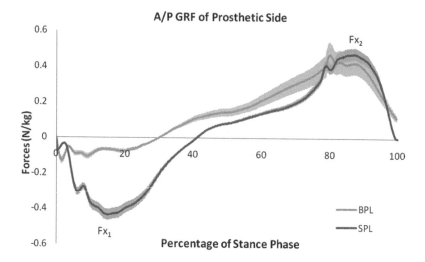

**FIGURE 12.11** A/P GRF of prosthetic side for one subject walking at self-selected pace using the BPL and SPL. Body weight was used to normalize GRFs (BW). The shaded field denotes the standard deviation.

limb (Powers et al., 1994; Snyder et al., 1995). In contrast to these findings, no important difference (p >0.05) in vertical GRFs was observed between the prosthetic and sound sides. The vertical GRFs is equal while walking with BPL or SPL (p >0.05).

A/P GRF values would indicate whether or not the prosthesis was correctly aligned. A/P GRF was found to be significantly different (p 0.05) in Fx1 of the SPL prosthetic hand, which had the highest significance among all breaking powers. This demonstrated the subjects' ability to monitor their breaking powers when wearing the SPL. Fx1 values for the prosthetic and sound sides of BPL is similar (p >0.05) to those for the sound side of SPL. There was no noticeable change in Fx2 (p >0.05) between the sound and prosthetic sides of BPL and SPL.

## 12.7.4 Range of Motions

Table 12.4 displays the typical hip, knee and ankle joint kinematics for all subjects. When participants wear either BPL or SPL, there were significant variations in maximal hip extension (p 0.05) on all sound sides. As opposed to other types, the magnitudes for the sound side of SPL were the largest (Figure 12.12). Subjects clearly felt more stable during gait when their body weights were accompanied by the sound leg.

Walking with BPL had no important difference (p >0.05) from walking with SPL in knee flexion and expansion of both limbs. The participants in this study walked with 15% less knee flexion and 10% more extension than the average person (Su et al., 2007). To minimize relative motion between the residual limb and the prosthetic socket, the subjects can reduce knee flexion. Figure 12.13 depicts the angular orientation of both limbs' knee joints while wearing BPL and SPL. For maximal dorsiflexion values, ankle joint angular values were equivalent (p >0.05) between BPL and SPL. For full plantar flexion, only the prosthetic side of SPL showed a significant

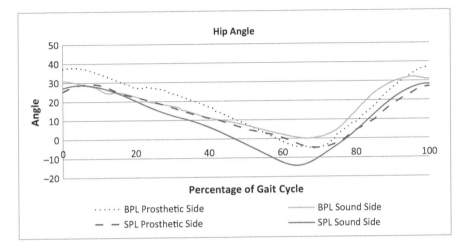

**FIGURE 12.12** The angular orientation of the hip joint in both subjects (N = 6). Positive values denote flexion, while negative values denote expansion.

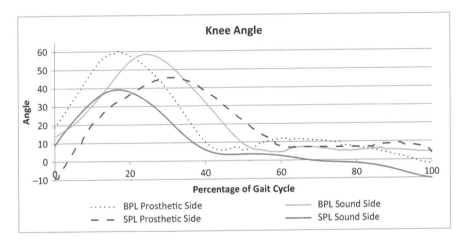

**FIGURE 12.13** For all participants (N = 6), the angular orientation of the knee joint was measured. Positive values denote flexion, while negative values denote expansion.

difference (p 0.05). Furthermore, of all limbs, the prosthetic side of SPL produced the highest magnitudes for both dorsi flexion and plantar flexion.

### 12.7.5 SOCKET INTERFACE PRESSURE ANALYSIS

This study's supporting evidence was the residual limb interface strain, which confirmed that no high stress pressure or discomfort existed on the subjects' residual limbs. Anterior, dorsal, medial and lateral sensor classes were used to conduct socket interface pressure tests. The sensor's length was divided into three boxes,

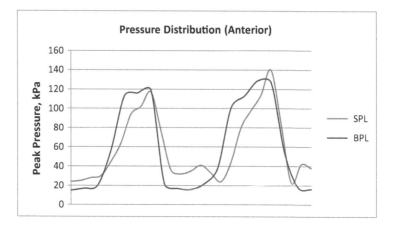

**FIGURE 12.14** When walking using BPL and SPL for two consecutive steps, the average peak distribution at the anterior region of residual limbs for six subjects was calculated.

representing the top, medial and distal regions of the respective areas. Figure 12.15 depicts the maximal pressure produced during S1's stance process when wearing BPL and SPL. The maximum pressure for each box is seen in the upper right corner of the corresponding box.

Figure 12.14 shows that the pressure interface was high on the anterior and medial areas, particularly for SPL, with values of 104 and 140kPa, respectively. These regions

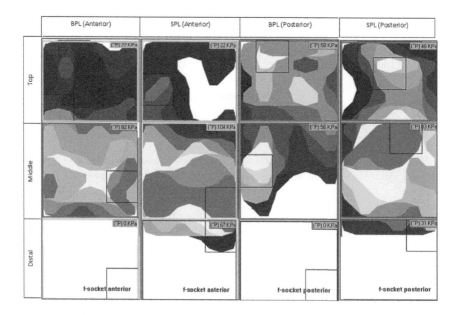

**FIGURE 12.15** Maximum stance pressures for BPL and SPL in the vicinity of the anterior and posterior residual limbs.

denote the patellar tendon and the tibial tuberosity, all of which are highly pressure resistant. According to Zhang and Lee (2006), the pain threshold for patellar tendon was 1.00MPa, while the pain threshold for tibial tuberosity was 0.89MPa. In their analysis, the lowest pain threshold measured was 0.45MPa at the distal end of the fibula.

## 12.8   CONCLUSIONS AND RECOMMENDATIONS

The bamboo pylon was subjected to three mechanical checks and was analyzed using a finite element model. According to the findings, bamboo is three times stronger than fibre-reinforced plastic and two times stronger than aluminium. Transtibial pylon materials such as fibre-reinforced plastic and aluminium are often used.

Bamboo, despite its resilience, had physical limitations in terms of water resistance and longevity. These constraints were resolved through the lamination process and oil treatment, which improved its anatomical structures. The overall cost of developing bamboo pylon was RM10, with just a few manufacturing processes needed. Aside from that, no assessment of bamboo pylon longevity was performed in this report. This assessment should be performed using a complex test under the Universal Testing Machine for further investigation.

A clinical evaluation was performed to ascertain the efficacy of the bamboo transtibial pylon in amputee gait. There was no important change in vertical and A/P GRF (p >0.05) between BPL and SPL. But for the hip extension (p 0.05), joint kinematics for BPL is equivalent to joint kinematics for SPL (p >0.05). Cadence, step length, and walking speed of BPL were comparable to SPL in spatio–temporal parameters (p >0.05).

## REFERENCES

Czerniecki, J.M., Gitter, A., & Weaver, K. (1994). Effect of alterations in prosthetic shank mass on the metabolic costs of ambulation in above-knee amputees. *Am J Phys Med Rehabil*, 73, 348–352.

Fu, J. (2001). *The Competitive Strength of Moso Bamboo in the Natural Mixed Evergreen Broad-Leaved Forests of the Fujian Province, China.* Doctoral Dissertation, Faculty of Forestry Science and Forest Ecology. Georg-August-Universität Göttingen, Germany.

Godfrey, C.M., Brett, R., & Jousse, A.T. (1977). Foot mass effect on gait in the prosthetic limb. *Arch Phys Med Rehabil*, 58, 268–269.

Hale, S.A. (1990). Analysis of the swing phase dynamics and muscular effort of the above-knee amputee for varying prosthetic shank loads. *Prosthet Orthot Int*, 14, 125–135.

Hahl, J., Taya, M., & Saito, M. (2000). Optimization of mass-produced trans-tibial prosthesis made of pultruded fiber reinforced plastic. *Materials Science and Engineering: A*, 285(1–2), 91–98.

Jain, S., Kumar, R., & Jindal, U.C. (1992). Mechanical behaviour of bamboo and bamboo composite, *J Mater Sci*, 27, 4,598–4,604.

Joachim, R. Harald, H., & Martin, B. (2007). *Mechanical behaviour of engineering materials: metals, ceramics, polymers and composites.* Springer. ISBN 978-3642092527.

Lee, A.W.C., Xuesong, B., & Perry, N.P. (1994). Selected physical and mechanical properties of giant timber bamboo grown in South Carolina. *Forest Prod J* 44(9), 40–46

Mathew, G., & Nair, K. S. S. (1990). *Storage pests of bamboos in Kerala.* In *Bamboos. Current research. Proceedings of the International Bamboo Workshop*, Cochin, India, 14–18 Nov. 1988. (pp. 212–214). Kerala Forest Research Institute, Kerala, India.

Meikle, B., Boulias, C., Pauley, T., & Devlin, M. (2003). Does increased prosthetic weight affect gait speed and patient preference in dysvascular transfemoral amputees? *Arch Phys Med Rehabil*, 84, 1,657–1,661.

Nghia, N.H. (2006). *Bamboos of Vietnam*. Agricultural Publishing House, Hanoi.

Powers, C.M., Torburn, L., Perry, J., & Ayyappa, E. (1994). Influence of prosthetic foot design on sound limb loading in adults with unilateral below-knee amputations. *Arch Phys Med Rehabil*. 75(7), 825–829.

Rho, J.Y. (1993). Young's modulus of trabecular and cortical bone material: ultrasonic and microtensile measurements. *J Biomech* 26(2), 111–119.

Shasmin, H. N., Osman, N. A., & Abd Latif, L. (2008). *Economical tube adapter material in below knee prosthesis*. In *4th Kuala Lumpur International Conference on Biomedical Engineering 2008* (pp. 407–409). Springer, Berlin, Heidelberg.

Shin, F.G., Xian, X.J., Zhena, W.P., & Yipp, M.W. (1989). Analysis of the mechanical properties and microstructure of bamboo epoxy composites, *J Mater Sci*, 24, 3,483–3,490.

Skinner, H.B. & Mote, C.D. (1989). Optimization of amputee prosthetic weight and weight distribution. *Rehabil Res Dev Prog Rep*, 26, 32–33.

Snyder, R.D., Powers, C.M., Fontaine, C., & Perry, J. (1995). The effect of five prosthetic feet on the gait and loading of the sound limb in dysvascular below-knee amputees. *J Rehabil Res Dev*, 32(4), 309–315.

Sulaiman, O., Hashim, R., Wahab, R., Ismail, Z.A., Samsi, H.W., & Mohamed, A. (2006). Evaluation of shear strength of oil treated laminated bamboo. *Bioresour Technol*, 97, 2,466–2,469.

Su, P., Gard, S.A., Lipschutz, R.D., & Kuiken, T.A. (2007). Gait characteristics of persons with bilateral transtibial amputations, *J Rehab Res Develop*, 44(4), 491–502.

Verterra, R.T.F. (2007). Stress-strain Diagram. Retrieved 20 February 2008 from http://www.mathalino.com/reviewer/mechanics-and-strength-of-materials/stress-strain-diagram.

Whittle, M.W. (1996). Ground reaction forces. *Gait analysis: An introduction*. Second Edition. Reed Educational and Professional Publishing Ltd. pp. 90–91.

Xiaobing, Y. (2007). Bamboo: Structure and Culture: Utilizing Bamboo in the Industrial Context with Reference to Its Structural and Cultural Dimensions. Doctoral Dissertation, Fachbereich Kunst und Design der Universität Duisburg-Essen, Germany.

Zhang, M. & Lee, W.C.C. (2006). Quantifying the regional load-bearing ability of transtibial stumps, *Prosthet Orthot Int*, 30(1), 25–34, doi:10.1080/03093640500468074.

# Index